T0360461

Consumer Society and Ecological Crisis

Consumer Society and Ecological Crisis advances a critique of consumer capitalism and its role in driving environmental degradation and climate crisis, placing a spotlight on how marketing and distribution activities help maintain unsustainable levels of consumption.

Rather than focusing on the most visible sites of promotional communication, Meier examines less conspicuous facets of marketing and logistics in distinct chapters on plastic packaging, e-commerce, and sustainability pledges in the fossil fuel sector. These three main chapters each explore links between ecological crisis and consumer capitalism, drawing on critical theory and Marxist thought. The topics of consumer convenience, speed, and economic growth – and the role of fossil fuels as guarantor of these logics of consumer society – unite the critical analysis.

Situated in the field of media and communication studies and adopting an interdisciplinary approach, this book will be of interest to academics, researchers, and students in the areas of media and communication studies, cultural studies, sociology, geography, philosophy, political science, and advertising.

Leslie M. Meier is Lecturer in Media and Communication at the University of Leeds, UK.

Routledge Critical Advertising Studies
Series Editor: Jonathan Hardy

Routledge Critical Advertising Studies tracks the profound changes that have taken place in the field of advertising. Presenting thought-provoking scholarship from both prominent scholars and emerging researchers, these ground-breaking short form publications cover cutting-edge research concerns and contemporary issues within the field. Titles in the series explore emerging trends, present detailed case studies and offer new assessments of topics such as branded content, economic surveillance, product placement, gender in marketing, and promotional screen media. Responding quickly to the latest developments in the field, the series is intellectually compelling, refreshingly open, provocative and action-oriented.

Branded Entertainment and Cinema
The Marketisation of Italian Film
Gloria Dagnino

Branding Diversity
New Advertising and Cultural Strategies
Susie Khamis

Branded Entertainment in Korea
Hyunsun Yoon

African Luxury Branding
From Soft Power to Queer Futures
Mehita Iqani

Consumer Society and Ecological Crisis
Leslie M. Meier

For more information about this series, please visit: www.routledge.com/ Routledge-Critical-Advertising-Studies/book-series/RCAS

Consumer Society and Ecological Crisis

Leslie M. Meier

Routledge
Taylor & Francis Group

LONDON AND NEW YORK

First published 2023
by Routledge
4 Park Square, Milton Park, Abingdon, Oxon OX14 4RN

and by Routledge
605 Third Avenue, New York, NY 10158

Routledge is an imprint of the Taylor & Francis Group, an informa business

British Library Cataloguing-in-Publication Data
A catalogue record for this book is available from the British Library

Library of Congress Cataloging-in-Publication Data
Names: Meier, Leslie M., author.
Title: Consumer society and ecological crisis / Leslie M. Meier.
Description: Abingdon, Oxon ; New York, NY : Routledge, 2023. |
    Series: Routledge critical advertising studies | Includes bibliographical
    references and index.
Identifiers: LCCN 2022042722 | ISBN 9780367431624 (hardback) |
    ISBN 9781032439389 (paperback) | ISBN 9781003001621 (ebook)
Subjects: LCSH: Consumption (Economics)—Environmental aspects. |
    Consumer behavior—Environmental aspects. | Environmentalism—
    Economic aspects.
Classification: LCC HC79.C6 M43 2023 | DDC 339.4/7—dc23/eng/
    20220908
LC record available at https://lccn.loc.gov/2022042722

ISBN: 978-0-367-43162-4 (hbk)
ISBN: 978-1-032-43938-9 (pbk)
ISBN: 978-1-003-00162-1 (ebk)

DOI: 10.4324/9781003001621

Typeset in Times New Roman
by Apex CoVantage, LLC

Contents

Acknowledgements

Writing this book was made possible by the support of many colleagues and friends. Jonathan Hardy was an exceedingly generous series editor, and I am very thankful for his advice and input on the manuscript. I am deeply grateful for the extensive written feedback from and many enriching conversations with Matt Stahl. At the University of Leeds, I would like to thank Bethany Klein and David Hesmondhalgh for their comments, discussion, and support. Thanks also to Benedetta Brevini and Vincent Manzerolle for providing productive feedback. I would like to extend my gratitude to the editorial team at Routledge: Margaret Farrelly, Elizabeth Cox, Priscille Biehlmann, and especially Hannah McKeating.

Additionally, I would like to acknowledge the generous period of research leave provided by the School of Media and Communication at the University of Leeds, which allowed me additional time to explore the ideas developed in this book. Thanks to members of the interdisciplinary Climate Justice Research Network at the University of Leeds. Our reading group sessions and research presentations have broadened my understanding of consumption, capitalism, and ecological crisis in powerful ways. I would like to express my gratitude to the Centre for Research in Communication and Culture at Loughborough University and the Department of Culture, Media & Creative Industries at King's College London for providing me with the opportunity to present chapters from the book as an invited research seminar speaker.

Finally, I would like to thank my family. I am especially grateful for the support of Judy Meier, Don Meier, Carol Bonli, and David Bonli. I miss Klaasje (Nicki) Soiseth, who showed me what it means to be a critical thinker, and Kathleen Meier, whose appreciation of nature and wildlife left a profound impression on me. Most of all, I thank Liv Bonli, my partner in everything.

Introduction

Promoting Consumption

Our consumer way of life, and the system of production and promotion that propels it, exist in a lopsided relation to the planet. Manufacturers treat raw materials and natural resources as an unending or self-replenishing stockpile to be plundered and transformed into commodities, rendered vehicles for capital accumulation. Advertisers inform us about and assign meanings to these commodities, mobilising financial and symbolic resources to promote the joys of consumption and the social significance of brands. The expansion of advertising has helped fuel 'an upward spiral of accumulation' among affluent populations, at the same time narrowing 'our society's central vision of the future – more of the same' (Lewis 2013: 71). The escalation of consumption is not the product of increased manufacturing and advertising alone. More mundane aspects of marketing and distribution have accelerated the circulation of physical commodities – strategies and activities that shape which goods end up on our shelves, how they get there, and the 'responsibility' of the corporations involved. The 'consumer' selects from what is on offer – commodities that are at once symbolic and material – to satisfy immediate needs, wants, and desires. Biodiversity loss, climate change, plastic pollution, and other ecological crises are the legacy of a system in which longer-term needs, wants, and desires, including environmental ones, are pushed from view.

Consumer Society and Ecological Crisis examines the contribution of marketing, broadly understood, and distribution to driving unsustainable rates of consumption. It asks students and researchers with an interest in advertising to take a step back from the immediate concerns and specific content of advertising practices and discourses. Instead, as part of a wider effort to 'green' media studies (Maxwell and Miller 2015), the following chapters invite critical reflection on: business practices that advertising works alongside in the consumer capitalist system; and

DOI: 10.4324/9781003001621-1

the environmental fallout of that system. After all, the basic purpose of advertising is:

> to get you to spend your money. . . . In a capitalist society, it is important that we all consume as much and as frequently as possible. This is what keeps the system functioning smoothly.
>
> (Holm 2017: 75)

Consumer capitalism is premised on expanding and accelerating the production and circulation of commodities in perpetuity. This book contributes to a growing literature on media, communication, and ecological crisis (Maxwell and Miller 2012; Lewis 2013; Maxwell, Raundalen, and Vestberg 2015; Cubitt 2017; Brevini and Murdock 2017; Park 2021; Aronczyk and Espinoza 2022; Brevini 2022), while attending to concerns about the commodity form, instrumental reason, and the domination of nature expressed in the consumer society critiques of the Frankfurt School (Adorno 2001 [1963]; Horkheimer and Adorno 2002 [1944]; Horkheimer 2004 [1947]; Adorno 2005 [1951]). In addition to media and communication studies, I will draw on fields ranging from sociology, geography, history, and philosophy to marketing, management, and logistics as I advance an environmental critique of contemporary consumer societies.

More specifically, I will examine disparate sectors and aspects of marketing and distribution that contribute to ecologically damaging forms of consumption: fast-moving consumer goods (FMCG) and plastic packaging; e-commerce and logistics; and fossil fuel corporations and their 'sustainability' promises. By advancing a critique of capitalism's 'predatory, extractive relation' to nature (Fraser 2021: 101), I will show the immense environmental price paid for a system in which '[t]he wealth of societies . . . appears as an "immense collection of commodities"' (Marx 1990 [1867]: 125). As we shall see in Chapter 1, plastic packaging allows for convenient and reliable delivery of food and drink products, cosmetics, and other 'disposable' consumer goods. Convenience in consumption comes at the cost of emissions and plastic pollution. Chapter 2 explores how the success of e-commerce is contingent on fossil-fuelled delivery. It is not the 'click of a button' that ultimately delivers the goods, but carbon-intensive freight transport and individualised courier-based delivery systems that contribute to the climate crisis. Digital exchange expedites physical distribution. Turning to corporate sustainability initiatives, Chapter 3 highlights how promising major changes later allows the fossil fuel industry to protect the bottom line and defer meaningful change now. 'Net zero' pledges reveal a temporal dimension to 'greenwashing', using future corporate behaviour to forge reputations for environmental responsibility in the present. The final chapter returns to key dimensions of consumption examined throughout the book and expands on the relevance of understanding the relationship between marketing, distribution, and capitalism to studies of advertising. These chapters offer different ways into understanding the damaging tempo of resource-intensive consumer societies. By placing a spotlight on the role played by behind-the-scenes or inconspicuous marketing and logistics activities, I hope to broaden understandings of the forces promoting over-consumption as part of an effort to stem it. The remainder of this intro-ductory chapter will discuss key terms and conceptual issues relevant to consumer society debates before providing an outline of the chapters.

Consumer Society, Capitalism, and Natural Resources

The concepts of *consumer society* and *capitalism* warrant explanation. The negative overtones associated with the former may suggest that environ-mental problems are the consumer's fault – that consumers are only inter-ested in the acquisition of more and more stuff, and simply do not care about the ecological crises we face. As Roberta Sassatelli (2007: 2) points out:

> underneath the apparent simplicity of the expression 'consumer soci-ety' lies profound ambiguity. From its very first appearance, this term has been used more to convey condemnation than to describe; in par-ticular, instead of being deployed to comprehend what characterized actual consumer practices, it served to stigmatize what appeared to be a growing and uncontrolled passion for material things.

In my use, consumer society is a shorthand for consumer capitalist society. It is not intended to convey criticism of individuals, or to dismiss the com-plexity and meaning tied to consumption, but rather to designate a society characterised by growing consumption. Under capitalism, we all must con-sume commodities to live – to maintain our capacity to work and to take care of ourselves and our families. Large business corporations – the real power brokers whose decisions shape consumption – are obliged to expand production and consumption in a capitalist system that sets the pursuit of profits as the purpose of commodity production.

Capitalism is not simply an economic system. As David Harvey explains, we can understand capitalism as:

> any social formation in which processes of capital circulation and accu-mulation are hegemonic and dominant in providing and shaping the material, social and intellectual bases for social life.

(Harvey 2015: 7)

Capitalism requires and encompasses non-economic conditions and processes, as Fraser convincingly argues:

> More than a way of organizing economic production and exchange, it is also a way of organizing the relation of production and exchange to their *non-economic conditions of possibility*. . . . What is often overlooked . . . is that this [economic] realm is constitutively dependent – one could say, parasitic – on a host of social activities, political capacities, and natural processes that are defined in capitalist societies as non-economic.
>
> (Fraser 2021: 99; emphasis in original)

Unwaged domestic activities overwhelmingly undertaken by women and, especially relevant here, easy access to the planet's natural resources make capitalist production and exchange possible. Viewed in this way, we can recognise the tight relationship between capitalist productivity and fossil fuel dependence; nineteenth- and early twentieth-century capitalism were powered by coal, and the decades since largely have been powered by oil (Fraser 2021: 112–118). Capitalism not only involves the exploitation of workers but also the appropriation of 'Cheap Nature', as '[a]ppropriated nature becomes a productive force' (Moore 2015: 16).

Expanding production and consumption in the post-World War II period has left an especially powerful imprint on Earth's ecosystems, atmosphere, and geologic record. Within earth sciences research, the 'Great Acceleration' thesis links increasing pressure placed on the earth system to rising economic growth and resource use since 1950; dramatic spikes in primary energy use, water use, motor vehicle use, fertiliser consumption, real gross domestic product (GDP), and other socio-economic factors find counterparts in rising levels of greenhouse gases (carbon dioxide, nitrous oxide, and methane), ocean acidification, surface temperature, and other dimensions of the earth system (Steffen et al. 2015). Economic growth has driven material growth, or 'the increase in the quantity of matter and energy transformed by human societies (e.g. trees cut down, coal burned, plants and animals eaten)' (Kallis et al. 2020: 8). While some economists claim that economic growth can be 'decoupled' from rising resource use and environmental strain,

> [t]here is no good empirical evidence for the belief that the level of decoupling of increasing emissions and growth in GDP is likely within the time frames required to avoid dangerous climate change.
>
> (O'Neill 2018: 142)

Regardless of what might be possible, serious damage has already been done:

> the scale of the human economy has become so large that its everyday activities, such as carbon dioxide emissions and freshwater use, now threaten the fundamental biogeochemical processes of the planet.
>
> (Foster and Clark 2020: 246)

Economic success has bred environmental devastation.

The impact has been so significant that a new geologic era has been proposed to characterise the (variously periodised) age of humans: the Anthropocene (Crutzen and Stoermer 2000). According to Crutzen and Stoermer (2000: 17), it is 'during the past two centuries [that] the global effects of human activities have become clearly noticeable'. Critics of capitalism point out that these environmental impacts derive not from human activities in general, but rather from particular sorts of activities and a specific configuration of power:

> Climate change, species extinction, and ocean acidification are just some of the markers of what scientists call the Anthropocene, a geological period characterized by a dominant human influence on the functioning of the ecosystem. At the same time, it is important to problematize who the "we" are because the dangerous ecological conditions we *all* face today are the product of particular political and economic policies and practices aimed at exploiting nature for the benefit of *a few*.
>
> (Ergene, Banerjee, and Hoffman 2021: 1320; emphasis in original)

Given the dominant influence not of humanity per se but of the capitalist system and capitalists, the Capitalocene has been proposed as an alternative moniker for 'the historical era shaped by relations privileging the endless accumulation of capital' (Moore 2015: 173).

A remarkably small number of organisations are responsible for an extraordinary share of environmental damage. According to Richard Heede's (2014) research, 63 percent of historical industrial carbon dioxide and methane emissions can be attributed to just 90 'carbon majors' (comprising investor and state-owned organisations and nation states). This responsibility derives from these entities' 'operations and from products each company has extracted and supplied to global consumers who used these fuels as intended thereby releasing carbon dioxide (CO_2)' (Kenner and Heede 2021: 1). During a relatively short expanse of time, the planet has been repurposed and reimagined as an engine of capital accumulation and circulation. Fossil fuels have animated this system, expediting the manufacture, transport, and exchange of commodities.

Consumer capitalism relies on natural resources and is governed by powerful corporations. To achieve rates of consumption commensurate with production capacity, corporations draw on marketing and distribution activities ranging from advertising and public relations to packaging and logistics. As vital complements to fossil power, such activities have kept goods selling and moving *at accelerating rates*, translating more production into more consumption.

Marketing, Logistics, Consumption, and Waste

The circulation of commodities involves managing the circulation of meaning *and* material flows, both of which involve marketing activities. Marketing plays an essential role in creating conditions for consumption by encouraging the exchange and acquisition of commodities. According to business historian Richard S. Tedlow (1990: 5):

> Marketing is an elusive subject, difficult to discuss because it is difficult to define. It encompasses a wide range of activities from the technicalities of logistics to the purest speculation about what people want now, what they will want in the future, and how much they will pay to have their wants satisfied.

A conventional view on what is called the 'marketing mix' uses the 'four Ps' as a short-hand for classifying marketing activities: *product* (everything from design to brand name to packaging); *price* (the list price and credit and discount schemes); *promotion* (advertising, public relations, direct marketing, sales, and other promotions); and *place* (distribution strategies, transportation, inventory management, and various logistical considerations) (Armstrong, Kotler, and Opresnik 2016: 81).[1] While advertising is a familiar form of marketing, logistics and packaging are perhaps less so. I will discuss them in turn. Drawing attention to logistics and packaging reveals how consumption is contingent on various infrastructures, materials, and processes.

For analytic precision, we can separate out *logistics* as a distinct business activity, as it is not focused on persuasion but instead on the coordination of movement. Logistics is 'the management of the circulation of goods, materials, and related information' (Danyluk 2018: 630). This business function encompasses 'transport, storage and handling of products as they move from raw material source, through the production system to their final point of sale' or acquisition (McKinnon 2015: 3). As Martin

Christopher (2016: 28), an authority on logistics and supply chain management, acknowledges:

> Even though textbooks describe marketing as the management of the 'Four Ps' – product, price, promotion and place – it is probably true to say that, in practice, most of the emphasis has always been on the first three. 'Place', which might better be described in the words of the old cliché, 'the right product, in the right place at the right time', was rarely considered part of mainstream marketing.

In advanced capitalist societies, escalating consumer expectations have underlined the significance of logistics technologies and systems. 'Availability' – 'is the product in stock, can I have it now?' – has become a key customer consideration alongside price and brand name (Christopher 2016: 28). With e-commerce, logistics is tasked with fulfilling the promise of immediacy promoted by the digital storefront.

Attending to these systems of information and transport powerfully reveals how digital exchange is conditional upon material consumption. Logistics brings together software and infrastructures (e.g., roads, railways) in service of 'the movement of labor, commodities, and data across global supply chains' (Rossiter 2016: xv). In recent decades, a 'logistics revolution' has taken hold:

> By enhancing the mobility of both commodity capital and the production process itself, advances in logistics have been an essential, albeit neglected, condition of economic globalization since the 1970s. To foreground the role of logistics in recent processes of global economic integration is therefore to shift the emphasis . . . from instantaneous financial transactions and digital data transmission to the decidedly slower and bulkier flows of goods and materials that, even in today's "information age," sustain human populations, fuel urban growth, and structure the uneven conditions of everyday life.
>
> (Danyluk 2018: 632)

Resource-intensive expectations of convenience and speed of product delivery translate into a dangerous reliance on fossil fuels. In import-dependent countries such as Britain, for instance, vast amounts of energy are consumed in order to transport 'stuff' via freight systems (MacKay 2009: 90–94). Through e-commerce, growing reliance on couriers intensifies and individualises this dynamic. According to Alan McKinnon (2015: 3), the 'dominant paradigm' of logistics has involved 'organiz[ing] logistics in a way that maximizes profitability' without accounting for environmental costs, but this has started to change, given the substantial carbon dioxide emissions generated by freight transport and warehousing, and, hence, growing public and government pressure. Environmental damage can translate into reputational damage.

Packaging brings together the persuasive power of symbolic expression with seemingly more mundane considerations related to logistics and transportation. The package is designed to catch consumers' attention in retail contexts *and* support fast and convenient movement through distribution channels. As explained in a marketing textbook,

> Traditionally, the primary function of the package was to hold and protect the product. In recent times, however, packaging has become an important marketing tool as well. Increased competition and clutter on retail store shelves means that packages must now perform many sales tasks – from attracting buyers to communicating brand positioning to closing the sale.
>
> (Armstrong, Kotler, and Opresnik 2016: 238)

As a medium of both promotional communication and distribution, the package acts as 'the interface between the product and the consumer' (Olsson and Györei 2002: 232).

In recent decades, plastic has emerged as a preferred material among packaged goods manufacturers, helping deliver and display items as important as food and beverages. This shift has been a boon to the petrochemical industry but a disaster for the biophysical environment. Plastics manufacture is a very carbon intensive process:

> The petrochemical industry makes plastics from raw material 'feedstocks', which are derived from fossil fuels and other hydrocarbons through a process known as 'cracking', applying heat and pressure to break down heavy hydrocarbons into lighter molecules. Petrochemical expansion relies on (1) access to cheap and abundant 'virgin' (fossil fuel-based) feedstocks; and (2) continual growth in new plastics markets to absorb expanding production.
>
> (Mah 2022: 3)

The expansion of plastic packaging has created a plastic pollution crisis and yet more fossil fuel dependence.

The definition and understanding of *consumption* differ by discipline, and the term is used to refer to activities as varied as use, acquisition, and purchase. As Foster, Clark, and York (2010: 381) explain:

> In environmental terms, consumption means the using up of natural and physical resources. . . . In economics, in contrast, consumption is only

one part of aggregate economic demand – the part accounted for by the purchases of consumers.

By contrast, in sociologist Alan Warde's (2005: 137) conceptualisation·

> consumption [is] a process whereby agents engage in appropriation and appreciation, whether for utilitarian, expressive or contemplative purposes, of goods, services, performances, information or ambience, whether purchased or not, over which the agent has some degree of discretion.

In response to the perception that Marxist consumer society critiques placed too much emphasis on the power of production, consumption studies involving micro-level interpretation of habits and practices have highlighted agency in consumption (for an overview, see Schor 2007). Perhaps unsurprisingly, given the broad and rich range of activities characterised as consumption, a problematic sensibility developed in anthropology, David Graeber (2011: 490) points out, whereby it seemed that '[i]n denouncing consumption, we are denouncing what gives meaning to the lives of the very people we claim we wish to liberate'. Consumption has been conceptualised as an activity and form of communication that expresses identity and conveys social distinction, as a site of creativity and of manipulation, as an exercise of conformity and of liberation, and as something 'produced' by capitalists and as a form of resistance (for key debates, see Schor and Holt 2000; Featherstone 2007; Sassatelli 2007; Storey 2017). Thus, there is considerable academic disagreement regarding whether or not to direct critique towards consumption and a purported consumer society in the first place, with normative perspectives and critical priorities varying widely.

I will draw on a Marxian lens to conceptualise consumption, but I will not adopt a crude or deterministic approach. In a consumer society, people do not just consume anything: we consume commodities produced by capitalists primarily for the purpose of exchange, not meeting needs per se (although commodities certainly do meet needs as well). Under capitalism, the commodity form is 'bourgeois society['s] . . . economic cell-form' (Marx 1990 [1867]: 90). No endpoint can be reached at which societies have acquired enough commodities – when our needs and desires have been sufficiently met – as this would cause economic crisis. Instead, 'the law of value . . . commands maximum expansion of capital at maximum speed' (Fraser 2021: 121). Importantly, production and consumption constitute different moments in a common process, and '[e]ven Marx accepted that consumption incentivizes production, just as production creates new conditions of consumption' (Harvey 2021: 84). Circulation activities 'help intensify the feedback loops between consumers and producers' (Pitts 2015: 216). Attending to this entire process is necessary to grasp the full environmental cost of consumer capitalism. By foregrounding the role of marketing and distribution, this book shifts focus from the figure of the 'consumer' to moments and dynamics that keep commodities circulating, driving economic and material growth.

In an age of multiplying ecological crises, it has become apparent that many consumer practices to which people have become accustomed will not be sustainable in the long term and have already caused considerable environmental damage. From my perspective, the profound social injustices and environmental issues arising from ecological crises justify critical evaluation of different forms of consumption – even those experienced as meaningful. Often implied by the term consumption is 'overconsumption'. While also imperfect, the term nevertheless helpfully signals 'an excess variously and interconnectedly understood in terms of ecology, equity, well-being, social cohesion and morality' (Humphery 2010: 22).

Notions of overconsumption and excess express the unbalanced relation between consumer societies and biophysical environments. Opening up the analysis to questions about 'the *volume* of consumption' and 'the *distribution* of access' reveals the implications of individual consumption for the common good (Evans 2019: 511; emphasis in original). The capitalist system produces asymmetrical distributions of commodities, natural resources, and environmental consequences between the Global North and Global South, and between rich and poor. For instance, whereas '[t]he United States and Britain have each generated carbon emissions per capita since the Industrial Revolution that are ten times the global mean', Foster and Clark point out, '[t]he Global South is in many ways more immediately imperiled than the North' (2020: 259) – as are northern communities in the Arctic region. Within the United States and Britain, the richest one percent live especially high-carbon lifestyles, consuming luxuries (e.g., luxury cars, private jets, yachts) and adopting practices of 'hypermobility' (e.g., frequent driving and flying) that result in a severely disproportionate share of environmental damage (Kenner 2019: 16–22). The types, scale, speed, and justice of consumption are germane to environmental critiques of consumer society.

Using Warde's (2014: 281) conceptualisation of consumption as involving 'acquisition', 'appropriation', and 'appreciation' as a starting point, David M. Evans (2019: 506–507) helpfully proposes pairing these three As with three Ds – 'disposal', 'divestment', and 'devaluation' – when assessing the relationship between consumption and the environment. In this formulation, '*[a]cquisition* refers to processes of exchange and the ways in which people access the goods, services and experiences that they consume', and research on this aspect of consumption may investigate 'the political, economic and institutional arrangements that underpin the production and delivery, and the volume and distribution, of consumption' (Evans 2019: 506). Persuasive marketing combined with convenient distribution and exchange encourage the ongoing acquisition of commodities. These goods will eventually become rubbish requiring *disposal*. Thus, we can also attend to the political, economic, and institutional arrangements that govern the disposal of goods, or support 'second cycles of consumption' by others (Evans 2019: 507). As Evans (2019: 506–507) explains, appropriation has to do with meaning and 'what people do with goods, services and experiences', and *divestment* refers to how such 'attachments [can] be undone'. Finally, appreciation involves 'pleasure and satisfaction', including aesthetic considerations, and is counterposed with *devaluation* – when 'goods, services and experiences cease to operate effectively' or the 'loss of cultural meaning lead[s] to symbolic failure' (Evans 2019: 506–507). Commodities may transition from being useful to useless or less desirable for cultural or technical reasons (e.g., fashion, media, and information technology), or because they have already served their utilitarian function (e.g., commodity packaging).

As the production of new commodities increases, so too does the need to replace what is already owned, leading to criticisms of a 'throwaway society' (Packard 1961; for an overview, see Trentmann 2017: 622–675). The pervasiveness of 'planned obsolescence' underlines the wastefulness built into this system (Lewis 2013: 136–138; Foster and Clark 2020: 249; Park 2021: 120–122). Similarly, due to the acceleration of the 'fast-fashion cycle', rapid turnover in material goods rich in symbolic meaning – clothing, consumer electronics, furniture, and so on – has translated into 'an almost manic flow of goods through the cycle of purchase, use, and discard' (Schor 2013: 446). While consumption can be an important site of useful activity, social meaning, and cultural significance, even the social dimensions of consumption have a shelf life. Marketing influences what is seen as fashionable, desirable, and culturally relevant.

The notion of consumer sovereignty heralds 'individual choice' – which is problematically assumed to improve wellbeing (Tadajewski 2019: 205) – and consumer power. However, it is not as individuals but as a *collective force* that consumers can really exercise power and influence. As Kojin Karatani (2014: 193) points out, the capitalist system requires workers to serve a dual role as 'labourer-consumers':

> In order for the accumulation of capital to continue, it has to ceaselessly engage in the recruitment of new proletarians. Of course, they are simultaneously also new consumers. The participation of these new proletarian consumers is what makes possible the self-valorization of industrial capital.

Capitalists need labourer-consumers as customers. Hence, as a collective, labourer-consumers can exert pressure on manufacturers and marketers through organised critical feedback or even consumer boycotts.

However, capitalists likewise constitute a collective force, and they hold immense economic and social power. The marketing, distribution, and exchange of certain commodities and not others enables, compels, and forecloses certain forms of consumption, as consumers must choose from what is promoted and made available. Also, capitalists undertake environmentally consequential forms of consumption over which consumer influence is severely circumscribed.

'Productive' Consumption, Marketing, and the Realisation of Value

In a capitalist society, we may exercise choice in our consumption activities, but we must do so within the parameters of existing commercial structures, infrastructures, and social expectations, and the histories that shape their meaning and use. As Marx (1963 [1852]: 15) famously observed:

> Men make their own history, but they do not make it just as they please; they do not make it under circumstances chosen by themselves, but under circumstances directly encountered, given and transmitted from the past.

Consumers have little say over capitalist decisions regarding which commodities are produced in the first place, the materials and factory conditions shaping their production, the modes of transport by which they travel, the retail systems through which they are sold, and the energy sources and infrastructure powering this entire system. Infrastructures, systems, technologies, and practices inherited from past generations circumscribe the possibilities for change. Existing structures and processes 'lock in' a fixed range of consumer activities and habits, and 'high carbon systems have got into social life. . . . Systems both form and presuppose habits' (Urry 2011: 156). Many of these systems prioritise accommodating the circulation of capital over managing environmental impacts.

For Marx, *capital* is at once a thing and a process, as Harvey explains:

> [I]n Marx's definition, capital is constituted as *both* the process of circulation of value (a flow) *and* the stock of assets ("things" like commodities, money, production apparatus) implicated in those flows.
>
> (Harvey 1996: 50; emphasis in original; see also Harvey 2015: 73)

Logics of movement and increase are paramount. As Harvey's (2018) reading of Marx suggests, capital can be understood as 'value in motion', or as Foster and Clark (2020: 246) put it, 'self-expanding value' which 'is indistinguishable from the drive to accumulate on an ever-increasing scale.' This escalatory logic exists in tension with the planet's ability to replenish and recover:

> The raw materials and natural resources that keep the engines running are not inexhaustible. Quite to the contrary, they are inevitably limited. Even though many of them reproduce, they do not reproduce at the same (and increasing) pace at which we use them up. This is true for trees, fish, oil, etc. What's more, the natural environment *can* dispose of the waste and pollution modern societies produce, but it cannot do so at an ever increasing speed.
>
> (Rosa, Dörre, and Lessenich 2017: 62; emphasis in original)

An accumulation of environmental problems originates in the productiveness of capitalism.

Marx distinguishes 'productive consumption', defined as 'the production and sale of the use values that capital needs as means of production', from 'final consumption' (Harvey 2018: 12–13). The latter comprises:

> wage goods required by workers to reproduce themselves, luxury goods mainly if not entirely consumed by class factions within the bourgeoisie and the goods needed to sustain the state apparatus.
>
> (Harvey 2018: 13)

Consumer society critiques have focused disproportionately on final consumption, obscuring the social and environmental significance of productive consumption. To produce commodities destined for final consumption, capitalists must consume *means of production* commodities that support their production, which comprise:

> raw materials taken directly as free gifts from nature, partially finished products like auto parts or silicon chips, machines and the energy to power them, factories and the use of surrounding physical infrastructures (roads, sewers, water supplies, etc., which may be given free by the state or paid for collectively by many capitalists as well as other users).
>
> (Harvey 2018: 7–8)

In a similar vein, Foster, Clark, and York (2010: 382) distinguish 'investment goods' from 'consumption goods', with the former being 'specifically aimed at the expansion of this capacity to produce, and hence at the growth of the economy'. Capitalist investment is geared towards the creation of surpluses that support profit generation and reinvestment. Hence, commodity production begets more commodity production, driving economic and material growth. Failure to differentiate consumption by producers/investors and consumers/end users can give the impression:

> that the problem of the consumption of environmental resources is to be laid at the door of consumers alone. Yet to neglect in this way the impact of investors on the environment is to exclude the motor force of the capitalist economy.
>
> (Foster, Clark, and York 2010: 382)

Capitalists consume resources not only in the production but also the marketing and distribution of commodities.

Marketing plays a crucial role in making the investment pay off. The production of commodities in and of itself does not guarantee a sale, and successful exchange is required for value to be 'realised'. As Harvey (2018: 12) explains:

> The commodities are taken to market to be sold. In the course of a successful market transaction, value returns to its money form. . . . Marx calls this key transition in the value form 'the realisation of value'. But the metamorphosis that occurs when value is transformed from commodity to the money form may not go smoothly. If, for example, nobody wants, needs or desires a particular commodity then it has no value no matter how much labour time was expended in its production.

Companies that produce undesirable products go out of business. The massive advertising budgets dedicated to convincing consumers to buy commodities are a testament to capitalist concern about the realisation of value, as '[t]hings will not be used unless they sell. Things will not sell unless they are desirable in some way' (Pitts 2015: 209). Worldwide advertising spending was predicted to reach $657 billion in 2021 (Szalai 2021).

Marketers (i.e., the organisations, manufacturers, or brands that make use of marketing activities) assume both ideological and economic significance, not only filling the media with promotional messages but also providing the funding on which commercial media rely (Hardy 2014: 135). Marketing and logistics support the realisation of value by creating a perception of desirability and facilitating successful distribution and exchange:

Eliding or collapsing the distinction between production and realisation ignores the activities occurring between. Creative industries are only one part of this. Logistics, salesmanship and so on are others.

(Pitts 2015: 216)

Marketing and logistics strategies that accommodate the high-speed circulation of commodities also entrench fossil fuel consumption and dependence.

As Max Horkheimer and Theodor Adorno observe in their famous 'culture industry' chapter of *Dialectic of Enlightenment*:

the most powerful sectors of industry [are] steel, petroleum, electricity, chemicals. Compared to them the culture monopolies are weak and dependent. . . . The dependence of the most powerful broadcasting company on the electrical industry, or of film on the banks, characterizes the whole sphere, the individual sectors of which are themselves economically intertwined.

(2002 [1944]: 96)

The real holder of power over commercial media is the advertiser as financier. As we shall see, the fast-moving consumer goods, e-commerce, logistics, and oil and gas sectors are significant advertisers, and the petroleum and chemical industries have maintained their status as among the most powerful industries. Rather than focusing on the role of media in promoting overconsumption, this book emphasises additional ways these corporations keep commodities selling and moving, drawing on natural resources and generating carbon dioxide emissions in order to do so. It draws attention to how ecological crises derive from capitalism's instrumental relationship to the biophysical environment:

nature is today more than ever conceived as a mere tool of man. It is the object of total exploitation that has no aim set by reason, and therefore no limit.

(Horkheimer 2004 [1947]: 74)

Yet we clearly do face environmental limits and planetary boundaries.

Chapter Outline

In what follows, discrete but complementary chapters will illustrate problems and fundamental contradictions within the capitalist system that curb our ability to respond adequately to ecological crises. In Chapter 1, I will examine the contribution of fast-moving consumer goods (FMCG) to the plastic pollution crisis. Also called consumer packaged goods (CPG), these items comprise a range of products many of us use every day – from packaged food and drink to toiletries to cosmetics. Drawing on Nancy Fraser's (2021) concept of socio-ecological regimes of accumulation, I will provide an historical and political economic context for understanding the rise of plastic packaging. Extending Marx's (1990 [1867]) account of the commodity form, I will explain that with FMCG, we are actually dealing with two distinct commodities – the consumer good and single-use plastic packaging – and, hence, two sets of corporate interests: consumer goods and petrochemical companies. Together, these powerful sectors have used plastic to deliver environmentally damaging forms of consumer convenience.

Chapter 2 examines 'Big Retail' and 'Big Logistics' in the e-commerce sector. Online shopping accelerates the circulation of commodities via systems of personalised, fossil fuel-intensive delivery – a circulatory process I will conceptualise drawing on Harvey's (2018) reading of Marx and account of capital as 'value in motion'. Rather than focusing on the websites and applications that encourage purchases, I will call attention to the logistical coordination, courier services, and fossil fuel-centric infrastructures marshalled to bring about high-speed acquisition, driving unsustainable rates of consumption.

Chapter 3 turns to the fossil fuel industry on which both FMCG and e-commerce depend. It explores trends in business 'sustainability' through which fossil fuel corporations offer 'solutions' to ecological crises to which they contribute. Drawing on Fraser (2021) and Horkheimer and Adorno (2002 [1944]), I will explore the fundamental tension between ecological and economic sustainability. Under capitalism, climate change 'solutions' are constrained by the growth imperative. 'Net zero' pledges by fossil fuel corporations serve as a type of greenwashing by giving an impression of responsibility that serves as cover for yet more fossil fuel extraction now.

The concluding chapter returns to different ways of understanding consumption introduced above in relation to the cases of FMCG, e-commerce, Big Logistics, and fossil fuels, and discusses further the relevance of my account to advertising studies. The chapter summarises the critique advanced throughout the book: that the expanding and accelerating production and circulation of commodities required by capitalism necessarily produce ecological crises, and that marketing and logistics serve this end of high-speed commodity circulation. They are agents of acceleration. Powerful industries and corporations have a financial interest in preserving this capitalist system, and consumers have become accustomed to the conveniences provided by commodities such as disposable goods, online shopping, and fossil fuels. Substantial barriers, then, must be overcome to counter our environmentally damaging consumer way of life. It is my hope that drawing attention to these problems and their capitalist origins might spur dialogue and debate within and beyond media and communication studies about how we might imagine and create alternatives to resource-intensive consumption.

Note

1 Although rival schemas to the four Ps have been proposed, this break-down nevertheless helpfully identifies the broad range of activities that may be understood by marketers to fall under the umbrella of marketing.

References

Adorno, T. (2001 [1963]) *The culture industry: Selected essays on mass culture.* Edited by J.M. Bernstein. London: Routledge.

Adorno, T. (2005 [1951]) *Minima moralia: Reflections from damaged life.* Translated by E.F.N. Jephcott. New York: Verso.

Armstrong, G., Kotler, P. and Opresnik, M.O. (2016) *Marketing: An introduction.* 13th edn. Harlow, Essex: Pearson Education.

Aronczyk, A. and Espinoza, M.I. (2022) *A strategic nature: Public relations and the politics of American environmentalism.* New York: Oxford University Press.

Brevini, B. (2022) *Is AI good for the planet?* Cambridge: Polity.

Brevini, B. and Murdock, G. (eds.) (2017) *Carbon capitalism and communication: Confronting climate crisis.* Cham: Palgrave Macmillan.

Christopher, M. (2016) *Logistics and supply chain management.* 5th edn. Harlow: Pearson.

Crutzen, P.J. and Stoermer, E.F. (2000) 'The "Anthropocene"', *Global Change Newsletter,* 41(May), pp. 17–18.

Cubitt, S. (2017) *Finite media: Environmental implications of digital technologies.* Durham: Duke University Press.

Danyluk, M. (2018) 'Capital's logistical fix: Accumulation, globalization, and the survival of capitalism', *Environment and Planning D: Society and Space,* 36(4), pp. 630–647. doi:10.1177/0263775817703663

Ergene, S., Banerjee, S.B. and Hoffman, A.J. (2021) '(Un)Sustainability and organization studies: Towards a radical engagement', *Organization Studies,* 42(8), pp. 1319–1335. doi:10.1177/0170840620937892

Evans, D.M. (2019) 'What is consumption, where has it been going, and does it still matter?', *The Sociological Review,* 67(3), pp. 499–517. doi:10.1177/0038026118764028

Featherstone, M. (2007) *Consumer culture and postmodernism.* 2nd edn. London: SAGE.

Foster, J.B. and Clark, B. (2020) *The robbery of nature: Capitalism and the ecological rift.* New York: Monthly Review Press.

Foster, J.B, Clark, B. and York, R. (2010) *The ecological rift: Capitalism's war on the Earth.* New York: Monthly Review Press.

Fraser, N. (2021) 'Climates of capital: For a trans-environmental eco-socialism', *New Left Review',* 127(Jan/Feb), pp. 94–127.

Graeber, D. (2011) 'Consumption', *Current Anthropology*, 52(4), pp. 489–511.

Hardy, J. (2014) *Critical political economy of the media: An introduction*. London: Routledge.

Harvey, D. (1996) *Justice, nature and the geography of difference*. Malden, MA: Blackwell Publishing.

Harvey, D. (2015) *Seventeen contradictions and the end of capitalism*. London: Profile Books.

Harvey, D. (2018) *Marx, capital, and the madness of economic reason*. New York: Oxford University Press.

Harvey, D. (2021) 'Rate and mass: Perspectives from the *Grundrisse*', *New Left Review*, 130 (July/Aug), pp. 73–98.

Heede, R. (2014) 'Tracing anthropogenic carbon dioxide and methane emissions to fossil fuel and cement producers, 1854–2010', *Climatic Change*, 122(1), pp. 229–241. doi:10.1007/s10584-013-0986-y

Holm, N. (2017) *Advertising and consumer society: A critical introduction*. London: Palgrave Macmillan.

Horkheimer, M. (2004 [1947]) *Eclipse of reason*. London: Continuum.

Horkheimer, M. and Adorno, T. (2002 [1944]) *Dialectic of enlightenment: Philosophical fragments*. Translated by E. Jephcott. Stanford: Stanford University Press.

Humphery, K. (2010) *Excess: Anti-consumerism in the West*. Cambridge: Polity.

Kallis, G., Paulson, S., D'Alisa, G. and Demaria, F. (2020) *The case for degrowth*. Cambridge: Polity.

Karatani, K. (2014) *The structure of world history*. Durham, North Carolina: Duke University Press.

Kenner, D. (2019) *Carbon inequality: The role of the richest in climate change*. London: Routledge.

Kenner, D. and Heede, R. (2021) 'White knights, or horsemen of the apocalypse? Prospects for big oil to align emissions with a 1.5° C pathway', *Energy Research & Social Science*, 79. doi:10.1016/j.erss.2021.102049

Lewis, J. (2013) *Beyond consumer capitalism: Media and the limits to imagination*. Cambridge: Polity.

MacKay, D.J.C. (2009) *Sustainable energy – without the hot air*. Cambridge: UIT Cambridge.

Mah, A. (2022) *Plastic unlimited: How corporations are fuelling the ecological crisis and what we can do about it*. Cambridge: Polity.

Marx, K. (1963 [1852]) *The eighteenth brumaire of Louis Bonaparte*. New York: International Publishers.

Marx, K. (1990 [1867]) *Capital: A critique of political economy – Volume 1*. Translated by B. Fowkes. London: Penguin.

Maxwell, R. and Miller, T. (2012) *Greening the media*. Oxford: Oxford University Press.

Maxwell, R. and Miller, T. (2015) 'Greening media studies', in Maxwell, R., Raundalen, J. and Vestberg, N.L. (eds.) *Media and the ecological crisis*. London: Routledge, pp. 87–98.

Maxwell, R., Raundalen, J. and Vestberg, N.L. (eds.) (2015) *Media and the ecological crisis*. London: Routledge.

McKinnon, A. (2015) 'Environmental sustainability: A new priority for logistics managers', in McKinnon, A., Browne, M., Piecyk, M. and Whiteing, A. (eds.) *Green logistics: Improving the environmental sustainability of logistics*. 3rd edn. London: Kogan Page, pp. 3–31.

Moore, J.M. (2015) *Capitalism in the web of life: Ecology and the accumulation of capital*. London: Verso.

Olsson, A. and Györei, M. (2002) 'Packaging throughout the value chain in the customer perspective marketing mix', *Packaging Technology and Science: An International Journal*, 15(5), pp. 231–239. doi:10.1002/pts.585

O'Neill, J. (2018) 'How not to argue against growth: Happiness, austerity and inequality', in Rosa, H. and Henning, C. (eds.) *The good life beyond growth: New perspectives*. London: Routledge, pp. 141–152.

Packard, V. (1961) *The waste makers*. London: Longmans.

Park, D.J. (2021) *Media reform and the climate emergency: Rethinking communication in the struggle for a sustainable future*. Ann Arbor, Michigan: University of Michigan Press.

Pitts, F.H. (2015) 'Creative industries, value theory, and Michael Heinrich's new reading of Marx', *tripleC*, 13(1), pp. 192–222. doi:10.31269/triplec.v13i1.639

Rosa, H., Dörre, K. and Lessenich, S. (2017) 'Appropriation, activation and acceleration: The escalatory logics of capitalist modernity and the crises of dynamic stabilization', *Theory, Culture & Society*, 34(1), pp. 53–73. doi:10.1177/0263276416657600

Rossiter, N. (2016) *Software, infrastructure, labor: A media theory of logistical nightmares*. London: Routledge.

Sassatelli, R. (2007) *Consumer culture: History, theory and politics*. London: SAGE.

Schor, J.B. (2007) 'In defense of consumer critique: Revisiting the consumption debates of the twentieth century', *The ANNALS of the American Academy of Political and Social Science*, 611(May), pp. 16–30. doi:10.1177/0002716206299145

Schor, J.B. (2013) 'The paradox of materiality: Fashion, marketing and the planetary ecology', in West, E. and McAllister, M.P. (eds.) *The Routledge companion to advertising and promotional culture*. New York: Routledge, pp. 435–449.

Schor, J.B. and Holt, D. (eds.) (2000) *The consumer society reader*. New York: The New Press.

Steffen, W., Broadgate, W., Deutsch, L., Gaffney, O. and Ludwig, C. (2015) 'The trajectory of the Anthropocene: The great acceleration', *The Anthropocene Review*, 2(1), pp. 81–98. doi:10.1177/2053019614564785

Storey, J. (2017) *Theories of consumption*. London: Routledge.

Szalai, G. (2021) 'Global, U.S. ad spending to hit records on COVID rebound: forecast', *Hollywood Reporter*, 13 June. Available at: www.hollywoodreporter.com/business/business-news/advertising-record-2021-covid-rebound-1234966706/

Tadajewski, M. (2019) 'Critical reflections on the marketing concept and consumer sovereignty', in Tadajewski, M., Higgins, M., Denegri-Knott, J. and Varman, R. (eds.) *The Routledge companion to critical marketing*. London: Routledge, pp. 196–224.

Tedlow, R.S. (1990) *New and improved: The story of mass marketing in America*. Oxford: Heinemann Professional Publishing.

Trentmann, F. (2017) *Empire of things: How we became a world of consumers, from the fifteenth century to the twenty-first*. London: Penguin Books.

Urry, J. (2011) *Climate change and society*. Cambridge: Polity.

Warde, A. (2005) 'Consumption and theories of practice', *Journal of Consumer Culture*, 5(2), pp. 131–153. 10.1177/1469540505053090

Warde, A. (2014) 'After taste: Culture, consumption and theories of practice', *Journal of Consumer Culture*, 14(3), pp. 279–303. doi:10.1177/1469540514547828

1 Promoting Plastic
Short-lived Commodities, Long-term Waste

Introduction

Contemporary consumer societies have a serious plastic problem. It may seem like plastic has always been an essential accompaniment to consumer goods and, hence, an omnipresent feature of everyday life, yet the proliferation of non-military uses of plastic occurred only after World War II (Geyer, Jambeck, and Law 2017). As historian Susan Strasser (2019) observes:

> Even though many people alive today can remember a world without ubiquitous plastic, we still have a hard time remembering how we actually lived without it.

The plastics industry has been extraordinarily economically productive and ecologically destructive within its short history, with 'the total mass of plastics now exceed[ing] the total mass of all living mammals' (Carney Almroth in Carrington 2022). Forecasts predict continued growth in global plastic production in coming decades, even as we have already reached very concerning levels of plastic pollution (Persson et al. 2022).

This chapter explores the plastic pollution crisis and its relationship to consumer society. I will examine a product category labelled *fast-moving consumer goods* (FMCG) – a moniker that underlines its fast rate of sales, and, hence, rhythm of consumption and disposal. FMCG include heavily packaged foods, beverages, toiletries, and cosmetics. Greenpeace (2018: 10) named the FMCG sector 'a predominant force behind the throwaway economic model driving the plastic pollution crisis' – a model under which corporations use high sales volumes to counter-balance low profit margins. FMCG giants Coca-Cola, PepsiCo, Unilever, Nestlé, Procter & Gamble, and others have been named top polluters in plastic pollution brand audits (Break Free From Plastic 2021: 16). They are also top advertisers. This chapter will assess the character of packaging and FMCG as commodities

DOI: 10.4324/9781003001621-2

and sites of promotion and will attend to the capitalist relations underpinning their production and distribution.

The purpose of what follows is diagnostic. I will illuminate reasons why the plastic pollution crisis has proved so difficult to resolve, advancing a critique that builds on critical theory and Marxian thought. Drawing also on insights from marketing and business literature, media and communication studies, sociology, and the history of consumption, I will explore what the plastic package is *for* and who profits from it. Critical theorist Nancy Fraser (2021: 100) points out that in capitalist society, 'the system's economy is constitutively dependent on nature, both as a tap for production's inputs and as a sink for disposing its waste'. The plastic pollution crisis powerfully illustrates this dynamic of dependence. FMCG do produce benefits for consumers. These heavily packaged food and drink and personal care items are products for people short on time – people who consume in a hurry or on the go. However, I do not see consumer demand, simply understood, as the primary force behind this crisis. Consumers do not decide what ends up on shop shelves, how it is packaged, or how it is marketed to us. *Producers do.* FMCG companies' advertising fills our media, products fill our shelves, and plastic packaging pollutes our planet.

Promoting Convenience, Producing Waste

In order to understand how plastic has so rapidly come to dominate our marketplaces and our landscapes, it is necessary to consider who profits from plastic packaging, what purposes packaging serves, and what FMCG brands are really selling: convenience. Boxes and containers of all sorts are a feature of most consumer goods sectors. FMCG – items ranging from beverages to toothpaste to candy – are distinctive for their especially heavy reliance on packaging and significant contribution to the plastic pollution crisis. Coastal clean-ups have revealed an astonishing accumulation of plastic bottles and caps, food wrappers, plastic bags, straws, and plastic take-away containers alongside cigarette butts, glass and aluminium beverage containers, and other detritus (Ocean Conservancy 2021: 14). In this section, to start to get to the root of the problem, I will unpack the notion of convenience and its byproduct: waste.

Excessive packaging waste is not a new phenomenon and pre-dates the popularisation of plastic. The downsides of the marketer-cultivated 'throwaway spirit' were recognised by American social critic Vance Packard (1961) over sixty years ago. Rising levels of waste accompanied the post-World War II economic boom, as explained by Frank Trentmann (2017: 635–636), an historian and specialist in consumption:

Put bluntly, it had never been so cheap to buy new, nor so convenient to dispose of materials. As packaging and self-service spread, the concomitant rise in rubbish was stunning. In 1950 West Germany, peas, lentils and rice were mostly still sold loose. By the end of the decade, they all came pre-packaged. In the 1960s, household waste shot up from 200kg to 300kg a person a year, but, most worryingly, its volume doubled. In Berlin and Paris, half the rubbish was now packaging, mostly paper and cardboard; plastic still made up only 3 per cent of waste in 1971.

Plastic pollution has grown dramatically in the decades since.

The properties of plastic as a material are well-suited to the tasks performed by packaging, combining qualities of durability and disposability. In the 1970s, DuPont engineers discovered how to transform polyethylene terephthalate (PET) into bottles, creating a substitute for glass that was exceptionally light, transparent, and durable (Hawkins 2013: 53). Gay Hawkins (2013: 50; emphasis in original) observes that PET 'may have momentary functionality as packaging or as a container' but is 'something that is *made to be wasted*'. Industrial ecologist Roland Geyer, environmental engineer Jenna R. Jambeck, and oceanographer Kara Lavender Law (2017) point out that 'plastics' largest market is packaging, an application whose growth was accelerated by a global shift from reusable to single-use containers'. Between 1950 and 2015, an estimated 8.3 billion metric tons of virgin plastic and 6.3 billion metric tons of plastic waste were created (Geyer, Jambeck, and Law 2017).

Today, plastic and packaging waste are widely recognised as contributing to ecological crises and have become targets of media commentary and activist mobilisation. Single-use plastics reportedly:

> account for the majority of plastic thrown away the world over: more than 130 million metric tons in 2019 – almost all of which is burned, buried in landfill, or discarded directly into the environment.
>
> (Charles, Kimman, and Saran 2021: 19)

The discovery on a Hawaii shoreline of composite 'stones' featuring melted plastic, dubbed 'plastiglomerate', demonstrates how plastic has even had an impact on our geology (Corcoran, Moore, and Jazvac 2014).

Some governments have called for stricter regulatory measures – even outright bans – in efforts to tackle plastic pollution. The Government of Canada, for instance, initiated a ban covering some single-use plastic items to commence in December 2022: plastic carrier bags, stir sticks, six-pack rings, cutlery, straws, and difficult-to-recycle food packaging

(Lindeman 2022). What is exempt from Canada's ban – food wrappers, beverage containers, and packaged 'personal care' products (Flanagan 2020) – underlines a dilemma: FMCG product packaging, a major contributor to plastic pollution, is not the target of this ban. According to CTV News, reasons the Government of Canada provided 'for products not to be included in the ban include a lack of affordable and readily available alternatives' and 'the items in question serving an essential purpose' (Flanagan 2020). Banning six-pack rings and straws only requires surface-level adjustments. To ban plastic beverage and hair care bottles and food wrappers would require an overhaul of entire FMCG product categories and an undoing of the conveniences to which many are accustomed. Whether or not their purpose is 'essential' is highly debatable, but these items have become an entrenched part of everyday consumption for many.

FMCG are time-saving, short-lived commodities. Packaging enables convenient, effortless, and accelerated consumption – a stark contrast with bulk food shopping, for instance, which requires advance planning and bringing one's own reusable container, and does not cater to 'impulse' purchases. According to sociologist Elizabeth Shove (2003b: 410–411):

> The term convenience, originally referring to fitness for purpose, was adopted in the 1960s to describe arrangements, devices, or services that helped save or shift time; convenience food being the classic example. . . . All sorts of commodities are now sold as being convenient or as making life more convenient for those who use them.

From premade or quick preparation food products to kitchen appliances, the post-World War II food system, particularly in the United States, was managed and mediated by corporations promoting the virtues of convenience, 'a mentality they transmitted to customers through advertisements, marketing ploys, and other corporate efforts' (Weber 2020: 609). Growing reliance on packaged foods led to new ways of cooking, and the adoption of frozen and refrigerated foods altered shopping and storage practices, creating dependence on automobiles, refrigerators, and freezers (Twede 2016: 123) – energy-intensive technologies reliant on fossil fuel-dependent infrastructures.

The price paid for a food industry optimised for convenience has been steep, and it goes beyond packaging to encompass transport and other factors. This 'global food system [is] characterized by an increase in convenience and processed food', and in 2015 food systems accounted for one third of greenhouse gas emissions once production (including agriculture and land use), distribution, packaging, and consumption were considered

(Crippa et al. 2021: 200). FMCG giants such as Unilever and Nestlé contribute to this global food system (although not all FMCG are food or beverages).

Convenient consumption is coupled with convenient shopping. A comprehensive system of retailing through supermarkets, convenience stores, and the like renders fast-moving consumer goods available almost anytime, almost anywhere. 'Convenience products' (e.g., candy and fast food) are typically purchased 'frequently, immediately, and with minimal comparison and buying effort', according to a marketing textbook: 'Convenience products are usually low priced, and marketers place them in many locations to make them readily available when customers need or want them' (Armstrong, Kotler, and Opresnik 2016: 233). It is difficult to imagine a world in which shops do not feature huge refrigerators filled with beverages bottled in plastic, displays overflowing with crisps and chocolate bars in plastic wrappers, and shelves populated by shampoos and deodorants encased in plastic containers. An entire system of production and retail supports a convenience-oriented way of life.

Although premised on notions of consumer choice, shopping based on the common denominator of convenience arguably represents manufacturer and retailer choice and narrows the range of genuine alternatives considered. As Tim Wu (2018) opines:

> [c]onvenience seems to make our decisions for us, trumping what we like to imagine are our true preferences. (I prefer to brew my coffee, but Starbucks instant is so convenient I hardly ever do what I "prefer.") Easy is better, easiest is best. Convenience has the ability to make other options unthinkable.

Corporations perpetuate the notion that convenience is a default expectation.

Convenience is especially desirable under social conditions marked by what Hartmut Rosa (2013) terms 'social acceleration': in modern societies, new technologies and production techniques, increasing social change, and a fast pace of life come together in a mutually reinforcing dynamic of acceleration. Different consumption practices may be possible (e.g., buying unpackaged items from bulk food aisles and shops, making meals from scratch). However, under a capitalist system that demands ever more productivity of its workers, the ability to take the time needed to commit to such choices can be a luxury. An accelerating pace of life means that '[n]o matter how fast they run, . . . [individuals] almost never succeed in working off their *to do lists*. . . . [T]hey are indebted temporally' (Rosa 2017: 438; emphasis in original). These historically specific structural conditions, I suggest, provide a more compelling explanation for growing reliance on convenience products than simple consumer preference.

The FMCG sector has produced and responded to the dependence on convenience. As journalist Tim Dickinson (2020) observes:

> Over the past 70 years, we've gotten hooked on disposable goods and packaging – as plastics became the lifeblood of an American culture of speed, convenience, and disposability that's conquered the globe.

Rather than blaming the consumer, we ought to recognise the influence of powerful industries that profit from plastic. As Alice Mah (2022: 23) points out in *Plastic Unlimited*:

> Plastic is essential to modern life, from computers to washing machines to food supplies and medical equipment. . . . Yet most plastic is not essential. Nor does it have to be. To disentangle ourselves from toxic and wasteful plastics, we need to retrace the steps of our entanglement, starting with the corporations.

A 'throwaway culture' accommodates the interests of producers and capital-ism's drive for 'ever-faster circulation of commodities in order to increase sales. . . . More than 150 billion single-use beverage containers are pur-chased in the United States every year' (Foster and Clark 2020: 254–255). The planet, meanwhile, must absorb the never-ending accumulation of waste that results. Indeed, to grasp the structural forces contributing to the plastic pollution crisis, it is necessary to understand the assumptions and imperatives of the capitalist system of commodity production.

Moving Commodities, Promoting Exchange

Commodities typically do not circulate unaccompanied but instead are packaged. As Susan Willis (1991: 1) remarked in the early 1990s:

> Late twentieth-century commodity production has generated a compan-ion production of commodity packaging that is so much a part of the commodity form itself as to be one of the most unremarked features of daily life.

With FMCG, a primary commodity (the beverage, the cosmetics) is depen-dent on a secondary commodity (the container). In the current context (where unpackaged bulk retail is unavailable or too costly in terms of time), these two distinct commodities are largely inseparable, producing signifi-cant challenges to doing away with packaging altogether. With plastic pack-aging, the interests of industrial-scale plastic *consumers* (FMCG brands) and *producers* (petrochemical companies) align: 'Big Plastic isn't a single entity. It's more like a corporate supergroup: Big Oil meets Big Soda' (Dickinson 2020). More precisely, as Mah (2022: 15) explains:

> the plastics value chain span[s] a wide range of interconnected industries, from fossil fuels (oil, gas, and coal) and biofuels (sugar and biomass), to petrochemicals (transforming hydrocarbons into plastic resins), plastics converters (converting plastic resins into packaging and other end uses), plastic end markets (e.g., food and beverage companies and other 'fast-moving consumer goods' or FMCG companies), and waste management and recycling.

My interest here is in how the plastic pollution crisis has been propelled by FMCG brands and petrochemical companies (and their financiers) (Break Free From Plastic 2021; Greenpeace 2018; Charles, Kimman, and Saran 2021), as both sectors attempt to grow sales of their own core commodities.

According to Karl Marx, much can be learned about the capitalist mode of production and its social relations by analysing the commodity form. Commodities, the building blocks of the capitalist system, have a dual character as 'at the same time objects of utility and bearers of value' (Marx 1990 [1867]: 138). Marx (1990 [1867]: 125) terms the former the commodity's use value, satisfying people's needs, suggesting that '[t]he nature of these needs, whether they arise, for example, from the stomach, or the imagination, makes no difference'. For the consumer, FMCG may cater to the stomach *and* the imagination, insofar as the latter helps us choose which of the many beverages or snacks we purchase while responding to social desires to meet culturally defined trends and expectations of youthfulness, attractiveness, and so forth. The use value of packaging, on the other hand, relates to convenience and the ways it supports the distribution and exchange of consumer goods. Packaging is both a logistical and promotional medium (Prendergast and Pitt 1996). It enables safe movement from factories to retail sites, often across long distances, and offers up vital real estate – space for the inclusion of logos, images, and product descriptions – on which companies promote their brands. The package acts as 'a device for hailing the consumer and cueing his or her attention, by the use of color and design, to a particular brand-name commodity' (Willis 1991: 1). With fast-moving consumer goods, only one of these commodities is actually consumed in terms of being used up: the consumer good (food, beverages, toiletries, cosmetics). After consumption, the package remains as waste, its primary functions – transporting and selling consumer goods – already served. The vital differences in the material character of these two commodities becomes most obvious *after* exchange. Achieving successful exchange is the focus of the capitalist.

Under capitalism, a commodity's use value is subordinate to its exchange value. This latter dimension of the commodity refers to 'the quantitative relation, the proportion, in which use-values of one kind exchange for use-values of another kind' (Marx 1990 [1867]: 126). From the standpoint of the capitalist and capitalism, exchange is the main purpose of commodities. Commodity production is calibrated to creating profits, with a cycle of reinvestment allowing for more production, more exchange, and more profits. FMCG companies consume (purchase and use) another commodity, packaging, in order to realise profits from their consumer goods. Plastic manufacturers, on the other hand, produce plastic to realise profits from its various commodity applications.

Commodities may be *for* exchange, but they cannot sell themselves, and thereby realise their own value. In Marx's (1990 [1867]: 178) words, '[c]ommodities cannot themselves go to market and perform exchanges in their own right'. The logistical and promotional functions of packaging enable the realisation of value. The varied roles of packaging include 'protect[ing] the product in movement', 'attracting attention to a product and reinforcing a product's image', and 'provid[ing] convenience (for both middlemen and consumers) of handling and storing the product' (Prendergast and Pitt 1996). Once on the shelves of our supermarkets, health and beauty retailers, and convenience stores, the branded package is used to set the product apart from its competitors:

> Not every customer will see a brand's advertising, social media pages, or other promotions. However, all consumers who buy and use a product will interact regularly with its packaging.
>
> (Armstrong, Kotler, and Opresnik 2016: 238)

According to management scholars Christopher Simms and Paul Trott (2010: 398, 411), '[p]ackaging is of greatest significance as a marketing tool within the Fast Moving Consumer Goods (FMCG) industry' – a sector whose products 'are characterized by a high level of packaging that is inseparable from the core benefit/product itself'.

Plastic helps sell and move products far, fast, and wastefully. For instance, in addition to being 'exceptionally light, strong, and flexible . . . the PET bottle could also be freely shaped into an enormous range of styles to enhance shelf appeal' (Hawkins, Potter, and Race 2015: 15–16). Plastic is a material whose '[d]urability makes commodities mobile' and also accommodates 'long periods of storage and extensive supply chain connections', but whose disposability means it is 'always in the process of becoming waste' (Hawkins 2013: 57).

FMCG as a product category is notable not only for the significance of its packaging as a branding device, but also for the financial commitment top brands make to advertising (Sinclair 2012: 16, 24). Global corporations such as Procter & Gamble, Unilever, and Nestlé 'are perennially among the largest advertisers in all the major national markets' (Sinclair 2020: 7). Indeed, while Amazon displaced Procter & Gamble as the globe's top spender according to Ad Age's 2020 ranking, this was only the second time this FMCG giant had not ranked number one for advertising and marketing spending since 1987 (when the rankings were introduced) (Johnson 2020). Procter & Gamble returned to the top spot the following year, spending an estimated $11.5 billion globally (Johnson 2021a). L'Oréal (a personal care giant), Unilever, and Nestlé were also in the top 10 (Johnson 2021b). Coca-Cola ranked 17th (Bonilla 2021) and PepsiCo also spends big.

In sectors marked by high levels of product standardisation (e.g., colas), advertisers work particularly hard to differentiate otherwise similar commodities, associating particular brand names not with the product's function but instead symbolic meanings (Goldman and Papson 1996: 22) – here meeting the needs of the imagination. After all, consumer products 'not only satisfy [more utilitarian] needs, but also serve as markers and communicators for interpersonal distinctions and self-expression' (Leiss et al. 2018: 2). Advertising shapes perceptions of what it might mean to drink Coca-Cola and not Pepsi, or to use Herbal Essences and not Pantene shampoo. Advertisers may strategically stoke worries and anxieties, warning of social consequences of not using mouthwashes and deodorants,[1] or they may strategically reassure us that concern is unnecessary when we do face serious problems. For instance, green advertising may communicate the message that 'if we consume properly the Earth can sustain a global economic order based on high levels of consumption' when a soberer reality suggests that 'the stimulation of desire by advertising can only result in both social and ecological catastrophe' (Papson 1992: 409). Advertising has bolstered growing levels of production and consumption, and 'has encouraged wealthy societies towards insatiability' (Lewis 2013: 71). While consumer desires may be limitless, the capacity of the planet to provide is not.

With FMCG, advertising works in dialogue with the package as a promotional form. The effect is to transform the commodity into a 'commodity-sign' – a 'composite entity' that integrates:

> the significations imparted to it through advertising and design. Advertising transfers meanings on to a product from the outside, through repeated imagistic association. Through design, on the other hand, that same signification is stamped on to it materially. The result is a

dual-character object, the *commodity-sign*, which functions in circulation both as an object-to-be-sold and as the bearer of a promotional message.

(Wernick 1991: 15–16; emphasis in original)

The FMCG commodity is at once a brand and a consumer good – an entity whose content is both symbolic and material.

Advertising and design arguably contribute to commodity fetishism, as social qualities and associations (e.g., glamour, coolness) are assigned even to generic, non-luxury goods like soaps and snacks. For Marx (1990 [1867]: 165), the 'fetishism' of commodities speaks to:

> the definite social relation between men [sic] themselves which assumes . . . the fantastic form of a relation between things. In order, therefore, to find an analogy we must take flight into the misty realm of religion. There the products of the human brain appear as autonomous figures endowed with a life of their own, which enter into relations both with each other and with the human race. So it is in the world of commodities with the products of men's hands.

Just as religious doctrine obscures the power of organised religion, advertising obscures the power of capitalism. The social relations of production, including the exploitation of labour, are veiled. The social and environmental costs of capitalism – the material impacts of producing, distributing, and disposing of FMCG – remain opaque as the joys of purchasing and using commodities assume centre stage. The FMCG capitalist's extractive relation to the biophysical environment – the unrelenting consumption of fossil fuels and various raw materials in service of commodity exchange – is obfuscated.

Commodity fetishism distracts from the emissions and pollution that commodity production leaves in its wake. As we peer into convenience store or supermarket refrigerators filled with ice-cold beverages, we likely do not see the complex system of manufacture, transportation, and resource consumption that makes this form of refreshment possible, and 'when we gaze at food packaging, few of us will visualize the crude oil and gas within it' (Marriott and Minio-Paluello 2013: 172). Plastic appears as a problem only *after* consumption, and it is not a problem the producer has to worry about:

> waste disposal and environmental protection are not part of consumer capitalism's remit, [so] these duties are either ignored or regulated and funded by public institutions (usually a level of government).

(Lewis 2013: 27–28)

The public bears the financial burden, and the people and planet pay the ecological price for environmental problems disavowed by corporations as 'externalities' in economic terms (Lewis 2013: 27).

The Rise of Consumer Goods Giants and Consumer Packaging

Top plastic polluters Coca-Cola, Pepsi, Unilever, Nestlé, and Procter & Gamble (Break Free From Plastic 2021: 16) all have origins that can be traced to the nineteenth century – an age when new technologies, marketing techniques, and distribution systems enabled the rise of mass markets. While packaging food and drink for the purposes of storage and transport is an ancient practice (e.g., clay vessels, woven baskets), it was the mechanisation of bottle, carton, and can manufacture in the late nineteenth century that led to the popularisation of standardised packages designed for the individual consumer and household – and a concomitant move away from bulk selling (Twede 2016: 115–117). A commercially advantageous relationship between two sectors developed:

> As consumer product goods (CPG) manufacturers grew and consolidated, so did the surrounding industry. By the late 1920s, an identifiable "packaging supply industry" began to emerge, deriving from the previous material-based industries that made glass bottles, metal cans and paperboard cartons.
>
> (Twede 2016: 120)

According to packaging historian Diana Twede (2012: 262), 'mass-produced packaging was the ammunition for the "profit through volume" revolution'. This distinctive approach – 'selling many units at low margins rather than few units at high margins' – 'demanded mass marketing' commensurate with new mass production capabilities (Tedlow 1990: 344). The profit through volume model, and its persisting heavy reliance on packaging, are key reasons why the FMCG sector is a major contributor to the plastic pollution crisis (Greenpeace 2018: 10). High volumes of goods produced and sold translate into high volumes of packaging waste.

Mass-produced packaging allowed producers to sell differentiated consumer brands, rather than generic goods, distributed via new transportation technologies and networks. Innovations in packaging manufacture and design 'were occurring alongside the development of railways and steamships, which meant that manufacturers could trade nationally and internationally with much greater ease' (Moor 2007: 18). The power of brands reinforced the power of producers. Manufacturers could exercise 'a new kind of control' vis-à-vis retailers and wholesalers:

> No longer were customers to rely on the grocer's opinion about the best soap; no longer could wholesalers choose among various manufacturers who might fulfil their orders. People asked for Ivory, which could only be obtained from Procter and Gamble.
>
> (Strasser 1989: 30)

People asked for certain brands because of the influence of advertising, which contributed to the power of the manufacturer (Strasser 1989: 30; Willis 1991: 2).

Key industries consolidated around select groups of dominant corporations. In the United States, '[p]rior to the 1880s, most manufacturers were unknown to the people who bought their products', whereas after this time, a handful of successful companies, such as Campbell Soup, Carnation, Heinz, Pillsbury Flour, Procter & Gamble, and Quaker Oats, rose in familiarity (Tedlow 1990: 14). Early successes in sectors characterised by the 'profit-through-volume' approach granted such companies 'first-mover advantages' (Tedlow 1990: 345). By the middle of the twentieth century, a handful of FMCG giants controlled huge markets for their products. While by no means a perfect barometer, *Fortune* magazine's total revenue-based corporate rankings provide a rough measure of corporate dominance and, hence, power.[2] In the U.S. context, Procter & Gamble ranked 27th, Coca-Cola 126th, and PepsiCo 364th in the first edition of the Fortune 500, which was published in 1955 (CNN Money 2021a). By comparison, Exxon Mobil, Dupont, and Dow Chemical ranked second, 10th, and 69th (CNN Money 2021a). These oil and gas and chemical corporations are among the most dominant plastic producers today (Mah 2022: 16).

In the second half of the twentieth century, a configuration of power stabilised in which today's top plastic consumers were consistently significant FMCG revenue generators. In 1965, Procter & Gamble ranked 24th, Coca-Cola jumped to 68th, and PepsiCo trailed at 245th in the Fortune 500 (CNN Money 2021b). In the 1975 and 1985 editions, Procter & Gamble's position remained stable at 28th and 22nd, respectively, Coca-Cola ranked 73rd and 46th, respectively, and PepsiCo climbed from 89th to 40th, respectively (CNN Money 2021c; CNN Money 2021d). Finally, in 1995, we can see these American companies combined in a global list featuring European FMCG giants: Unilever ranked 38th, Nestlé 39th, Procter & Gamble 71st, PepsiCo 89th, and Coca-Cola 190th (Fortune 1996).

In the twenty-first century FMCG sector, corporations with roots in the nineteenth century dominate an oligopolistic market. In the United States, an investigation by *The Guardian* and Food and Water Watch revealed that '[n]inety-three per cent of the sodas we drink are owned by just three companies': Coca-Cola, Pepsi, and Keurig Dr Pepper, in that order (Lakhani, Uteuova, and Chang 2021). Nestlé, meanwhile, accounts for almost one quarter of the bottled water market (Lakhani, Uteuova, and Chang 2021). In the 2021 Global 500 (based on 2020 revenues), Nestlé ranked 79th, Procter & Gamble 128th, PepsiCo 131st, and Unilever 175th, with Coca-Cola falling to 370th (Fortune 2021). However, Coca-Cola was named the top plastic polluter, followed by these four FMCG giants, in Break Free from Plastic's 2021 global audit, which records 'branded plastic waste' (Break Free From Plastic 2021: 18).

Advertising and the promotion of brands act as a veneer that creates an illusion of market competition. By flooding the market with countless brands, FMCG corporations obscure this reality of corporate oligopoly under which a handful of giants dominate the food and snack, beverage, and personal care industries (Sinclair 2012: 22). According to John Sinclair (2012: 26):

> companies in oligopolistic markets tend to compete more in terms of advertising than price, which stabilises the market in their favour and keeps out prospective competitors.

Powerhouses Procter & Gamble and Unilever control 'hundreds of brands in several divisions, and in various parts of the world' (Sinclair 2012: 43). If you are seeking shampoo, you might grab a bottle of Aussie, Head & Shoulders, Herbal Essences, or Pantene – all of which are owned by Procter & Gamble (2022). Want some ice cream? Ben & Jerry's, Magnum, and Wall's are all Unilever (2022a) products. As highly diversified companies, Procter & Gamble and Unilever's brands cover a range of FMCG categories. Nestlé, whose stable of bottled water brands includes Nestlé Pure Life, Perrier, and San Pellegrino, is a giant in food and drink processing, owning brands ranging from Aero chocolate bars to Carnation evaporated milk to Nescafé coffee (Nestlé 2022). PepsiCo likewise has interests across the food and drink categories, including carbonated beverage brands Pepsi and Mountain Dew, orange juice brand Tropicana, and iced tea brands Lipton and Brisk, just to name a few (PepsiCo 2022). PepsiCo subsidiary Frito-Lay owns chip and snack brands Lay's, Doritos, Cheetos, and many others (Frito-Lay 2022). Beverage corporation the Coca-Cola Company's roster of 200 brands spans familiar carbonated beverages Coca-Cola, Sprite, and Fanta, bottled water brands Dasani and Glaceau waters, and juice brands Innocent, Minute Maid, and Simply (Coca-Cola Company 2022). Seemingly infinite varieties distract from this extraordinary concentration of power.

A handful of well-established corporations profit from volume-based business strategies. These corporations, which rely heavily on packaging, have a consequential role in shaping which goods end up on retailers' shelves, how they are distributed, and whether they are encased in plastic. Their brands are heavily advertised to consumers who ultimately must choose from what is on offer.

The Capitalist Growth Imperative, Fossil Fuels, and the 'Great Acceleration'

The rise of fast-moving consumer goods giants, proliferation of plastic packaging, and overall acceleration of commodity production was driven not simply by corporate power and the promotion of desire for commodities. Something vital is missing from my account so far. Explosive growth in capitalist economies and industries has been powered by the extraction, commercial exploitation, and burning of fossil fuels. Between 1850 and 1950, the consumption of coal and then oil and gas enabled expanding industrialisation and consumption rates in powerful capitalist economies (Murdock and Brevini 2017: 7). Consumption only accelerated thereafter:

> private consumption expenditures (the amount households spend on goods and services) increased more than fourfold from 1960 to 2000, even though the global population only doubled during this period.
>
> (Dauvergne 2008: 4)

Richard Heede's (2014) research rightly points to the contribution of the 'carbon majors' to historical emissions, but it is not only fossil fuel companies that have fuelled ecological crises. Those forms of commodity production, distribution, and exchange made possible by the consumption of fossil fuels, such as FMCG, have also made their imprint on the planet.

We can draw on research on the 'Anthropocene', or what Jason W. Moore (2015, 2016) calls instead the 'Capitalocene', to identify the impacts capitalism has had on earth systems.[3] While some periodisations date the Anthropocene to the Industrial Revolution, a 'Great Acceleration' in the post-World War II period brought about especially striking environmental impacts (Steffen et al. 2011: 849). As Will Steffen and his co-authors observe (Steffen et al. 2015: 94):

> In little over two generations – or a single lifetime – humanity (or until very recently a small fraction of it) has become a planetary-scale geological force. Hitherto human activities were insignificant compared with the biophysical Earth System, and the two could operate indepen-

dently. However, it is now impossible to view one as separate from the other.

Rises in surface air temperature, ocean acidification, and carbon dioxide, nitrous oxide, and methane emissions and other earth system trends accelerated after 1950 (Steffen et al. 2015: 87). A recent study conducted in southern California interprets the rise of plastic as yet more evidence of the post-1945 'Great Acceleration' within the Anthropocene, given the 'tightly coupled relationship between worldwide plastic production, regional population growth, and the plastic deposited in the sedimentary record' (Brandon, Jones, and Ohman 2019).

It is *capitalists*, not humanity in general, who have emerged as a geological and ecological force. As Graham Murdock and Benedetta Brevini (2017: 6) point out, 'the major contributors to greenhouse gases since 1750 have been the capitalist economies of Western Europe and North America'. Capitalism as a system 'drives global warming non-accidentally, by virtue of its very structure' (Fraser 2021: 98) – a structure dependent on perpetual growth.

Fraser's (2021) concept of 'socio-ecological regimes of accumulation' calls attention to the role of fossil fuels in capitalism's historical development. Such regimes are distinguished by 'a distinctive way of organizing the economy-nature relation. Each features characteristic methods of generating energy, extracting resources, and disposing of waste' (Fraser 2021: 109). Drawing on Fraser's terminology, we can locate the rise of packaged consumer goods during the regime of 'liberal-colonial capitalism' (nineteenth to early twentieth century) – a period during which:

> coal-fired steam powered the industrial revolution in production, [and] it also revolutionized transport. Railroads and steamships compressed space and quickened time, speeding the movement of raw materials and manufactures across great distances, thus accelerating capital's turnover and swelling profits.
>
> (Fraser 2021: 113)

Further acceleration was enabled during the oil-powered regime of 'state-managed capitalism' (the middle of the twentieth century), as the automobile – that 'icon of consumerist freedom, catalyst of highway construction, [and] enabler of suburbanization' (Fraser 2021: 115) – enabled new consumption practices. Driving to and from the mall or the supermarket, the consumer could accumulate a mounting number of branded products. The automobile allowed for greater individualisation, more frequency, and more volume in purchases on shopping trips, accelerating the pace and customising the time of shopping.

The age of oil and the car left an infrastructural inheritance: car-centric urban planning, highways, and parking lots. Forms of path dependence have kept fossil fuel consumption at the centre of capitalist economies in the decades since. As Elizabeth Shove (2003a: 12) explains:

> Initially used to explain why economically optimal forms of innovation are not always realized, the concept of path dependency highlights the extent to which existing technologies and practices structure avenues of future development.

Dynamics of path dependency '[m]ake reversals and dramatic changes of direction difficult' (Shove 2003a: 12). In our contemporary 'regime of financialized capitalism', industrial production in the Global South has been harnessed to enable overconsumption in the Global North, where 'steep rises in air travel, meat-eating, cement-making and overall material throughput' exemplify the entrenchment of carbon-intensive, environmentally destructive production and consumption (Fraser 2021: 118). FMCG freight transport relies on massive semi-trailers which, depending on the product, may require temperature control. In shops, plastic bottles are typically showcased in massive refrigerators not necessarily to avoid spoilage but instead so consumers can have a cold beverage *immediately*. Even technology innovators – e-commerce companies such as Amazon – rely on carbon-intensive infrastructures and modes of transport, including automobile-based couriers, to profit from data-driven marketing and personalisation and instantaneous digital transactions (see Chapter 2).

A succession of socio-ecological regimes of accumulation have added to the acceleration of production and consumption, and a corresponding acceleration of associated environmental impacts. Expanded capitalist infrastructures, faster and more distant commodity exchange, and inadequate restraints on polluting producers block alternatives to a carbon-intensive capitalist status quo. We are 'locked in'. The concept of 'carbon lock-in' refers to a type of path dependence:

> a combination of systematic forces . . . perpetuate fossil fuel-based infrastructures in spite of their known environmental externalities and the apparent existence of cost-neutral, or even cost-effective, remedies.
>
> (Unruh 2000: 817)

A system of fossil fuel-based technologies, infrastructures, and dominant institutions comes together to project carbon-intensive ways of doing things into the future.

Capitalism cannot avoid (let alone remedy) ecological crisis, in part because the capitalist imagination is defined by a blinkered instrumentalism: materials and environments are overwhelmingly rendered resources to be used in service of unrelenting commodity production and exchange. According to David Harvey (2015: 250):

> Nature is necessarily viewed by capital . . . as nothing more than a vast store of potential use values – as processes and things – that can be used directly or indirectly (through technologies) in the production and realisation of commodity values.

From this perspective, natural resources ranging from wood to fossil fuels as well as synthetic byproducts such as plastics *are for* exchange as commodities or potential commodities. Moreover, business success is contingent on growth and expansion: 'capital commands accumulation without end' (Fraser 2021: 100). Rosa, Dörre, and Lessenich's (2017: 60–61) conceptualisation of the 'dynamic stabilisation' to which modern capitalist societies are bound is instructive:

> [T]he principle of incessant increase, i.e. the logic of growth, augmentation and innovation . . . leads into a spiral of escalation. No matter how high the gross domestic product has been this year, it needs to be even higher next year, no matter how fast processes (for example in the production of goods and services) or the rates of innovation already are, they need to become even faster – for if they do not, there will be an economic slowdown, followed by a whole array of economic, social and political crises.

Ecological crises are a byproduct of efforts to avert economic, social, and political crises.

Through the capitalist lens, escalating capital accumulation and circulation can continue in perpetuity. However, the reality is that the biophysical world 'cannot really self-replenish without limit, [therefore] capitalism's economy is always on the verge of destabilizing its own ecological conditions of possibility' (Fraser 2021: 101). '[C]easeless expansion and constant motion' is enabled by the despoliation of ecosystems across the globe (Foster, Clark, and York 2010: 135–136). The 'predatory, extractive relation' between capital and nature (Fraser 2021: 101), combined with the short-term, instrumental bias of capitalist rationality, means that profits now trump environmental consequences later. '[D]amages are the flipside of the profits' (Fraser 2021: 100) – with the plastic pollution crisis offering a case in point.

Capitalism's growth imperative is agnostic as to the content – the industries and materials – of what is grown. Plastic is no exception. Roland

Barthes (1972: 97) remarked that, 'more than a substance, plastic is the very idea of its infinite transformation . . . [I]t is less a thing than the trace of a movement'. For Barthes (1972: 97), 'the quick-change artistry of plastic is absolute: it can become buckets as well as jewels'. However, the commodity form and growth imperative, not plastic's material possibilities, have defined its proliferation and use in capitalist societies. The plastics industry contributes to a diverse range of applications and products, including airplane and automobile parts, medical equipment, consumer electronics, furniture, and white goods, but packaging is a dominant application (British Plastics Federation 2016: 13–14). The British Plastics Federation (2022) characterises plastic packaging as 'a strong area of growth' and highlights how 'innovation in PET [is] pushing the boundaries of usage and ready meals and smart packaging [are] creating new opportunities'. Over half of global single-use plastic waste created in 2019 has been attributed to twenty polymer producers, led by American companies Exxon Mobil and Dow and Chinese company Sinopec (Charles, Kimman, and Saran 2021: 12). While plastic *could* become many things, it routinely *is* moulded in predictable and standardised ways by a handful of corporate powerhouses. Plastic pollution expresses a material 'trace' of the movement of commodities and capital, as the durability that was an asset when moving goods becomes a liability after consumption – a trace that resists decomposition.

The plastics industry pursues yet more growth. Ongoing investment in expanding infrastructures and facilities, such as ethane crackers, will support even greater plastic production (British Plastics Federation 2016: 36). As Harvey (2018: 150) points out, investments in 'fixed capital' facilities create a type of momentum that at once supports capital accumulation *and* 'imprison[s] [future production and consumption] within fixed ways of doing things. . . . The future is mortgaged to the past'. Committing to plastic production today tethers us to plastic consumption tomorrow, as '[p]lastic production is strongly associated with lock-in effects with raw materials, particularly fossil fuels' (Persson et al. 2022: 1514).

Conclusions and Closing Reflections

Consumer societies have burdened the biophysical environment and future generations with a staggering inheritance of waste. Plastic pollution is the material trace of the endlessly repeated moment when the commodity stops moving and value has been exhausted. This chapter has identified reasons why plastic is so prevalent and why reducing plastic packaging has proven so difficult. My explanation has hinged on dynamics fundamental to the commodity form, promotional communication, logistics, FMCG industry consolidation, corporate power, and the capitalist system and role of fossil fuels therein. Plastic's stubborn durability cannot be explained simply in terms of its qualities as a *material* (although these are of course important) but, crucially, its character as a disavowed *commodity*. The speed, convenience, and disposability of FMCG – qualities consumers have come to expect, in part, due to advertising and branding – are delivered by the companion commodity of plastic packaging. The disposability of the package is what delivers convenience.

FMCG and plastic producers benefit from continued, if not growing, production of plastic packaging. This logistical and promotional medium for FMCG corporations and profitable application for petrochemical companies aligns the interests of two powerful industries. In fact, for FMCG corporations, doing away with plastic packaging would require reimagining what the FMCG product is. What is a carbonated beverage or shampoo without its bottle, ice cream with no tub, or deodorant stick with no container? While alternative packaging materials can be and are being substituted and explored, as we have seen, plastic offers numerous advantages. Furthermore, options such as glass and aluminium are not without environmental impacts, as the plastics industry likes to point out.[4] As Greenpeace (2018: 33) sums up:

> On its own, material substitution simply shifts the burden of environmental impacts from one single-use material to another, without addressing the problems of overproduction and consumption. Increased material use, deforestation, land use, competition with food production, ocean pollution, recycling challenges . . . and high energy impacts may all be associated with other materials; companies must prevent replacing one problem with another.

Material substitution could help address the *plastic pollution* crisis, but it would not resolve the forms of environmental degradation that accompany the capitalist system.

It is not surprising that FMCG corporations are advocating recycling and the 'circular economy' as 'solutions' to the plastic pollution crisis:

> In contrast to climate change, the plastics crisis has not been met with corporate denial. The companies of Big Plastic are instead seeking to convince consumers and regulators that – despite having unleashed this torrent of pollution on the planet – they can be trusted to pioneer solutions that will make plastic use sustainable. They're touting a "circular economy," in which used plastic doesn't become waste but, instead, a feedstock for new products. A cynic might translate the concept into: Recycling, but for real this time.
>
> (Dickinson 2020)

Coca-Cola's response to Greenpeace's brand audit suggests that this company is banking on the power of recycling, stating that 'simply eliminating single-use plastic won't end ocean pollution' but instead that the answer lies in using recyclable materials, improving recycling collection systems and rates, '[e]ducating people about the importance of recycling and material reuse', and '[b]ringing the cost down and availability up for recycled materials' (Coca-Cola 2018). However, across the FMCG sector, we are starting to see a wide range and intensity of commitment to changing how products are packaged. Unilever (2022b), for instance, which aims to '[h]alve the amount of virgin plastic' in its packaging by 2025, is offering concentrated products that are less reliant on plastic and packaging and is supporting various refill and reuse options. At least on the surface, the major corporations appear to be on board for change. However, setting and meeting sustainability targets are altogether different.

By attending to the longer history of capitalism and its 'socio-ecological regimes' (Fraser 2021), we can discern a pattern. As Foster, Clark, and York (2010: 74) explain:

> one environmental crisis is "solved" (typically only in the short term) by changing the type of production process and generating a different crisis, such as how the shift from the use of wood to plastic in the manufacturing of many consumer goods replaced the problems associated with wood extraction with those associated with plastics production and disposal.

Regardless of the material used – aluminium, glass, wood – capitalism's growth imperative impels unsustainable escalation and expansion. Corporations cannot opt out of growth and succeed in business. The most decisive response to plastic pollution is resisted under capitalism: simply producing far less and, with some product categories, not at all. It may be that we do not actually need some FMCG, or at least that they are not worth the ecological cost. Moreover, initiatives intended to lessen environmental impacts in one area can distract from how such gains may be cancelled out by growth elsewhere.

The plastic pollution crisis is just one of multiple ecological crises, including the climate crisis. According to Trentmann (1017: 675):

> it is the busy, energy-hungry lives we lead at home and on the roads and in the air that are the real threat. . . . Recycling has been little more than a comforting distraction from the stuff that really matters.

I will turn to the sustainability pledges of the fossil fuel industry powering these carbon-intensive lives in Chapter 3. Next, in Chapter 2, I will explore how e-commerce has enabled a fossil-fuelled acceleration of production and consumption.

Notes

1 As Roland Marchand's (1985: 20) historical work on advertising shows, brands have tapped into 'social shame or personal fear' to promote such products since the 1920s.
2 The Fortune 500, which predates the Global 500, only includes U.S. companies. Therefore, European companies Unilever and Nestlé are absent in these early rankings.
3 In contrast to some, Moore stresses the environmental significance of early phases of capitalism, beginning with 'the long sixteenth century, ca. 1450–1640', noting the acceleration of large-scale deforestation, for example (Moore 2016: 94, 96–97).
4 For instance, as an article in plastics industry publication *Plastics Today* warns, 'switching from hundreds of millions of PET plastic bottles for drink products to hundreds of millions of glass bottles would result in an astounding amount of energy use just in terms of hot water and disinfectant cleaners to make reuse possible by sterilization. It would also result in product loss from breakage, energy used to transport truckloads of empty glass bottles from retail outlets, where consumers have returned them, to central washing locations and then shipped to bottling plants' (Goldsberry 2021).

References

Armstrong, G., Kotler, P. and Opresnik, M.O. (2016) *Marketing: An introduction.* 13th edn. Harlow, Essex: Pearson Education.

Barthes, R. (1972) *Mythologies.* Translated by A. Lavers. New York: Hill and Wang.

Bonilla, B. (2021) '12 biggest ad reviews of 2021: From Meta to Coca-Cola, this year saw some monumental agency shifts', *Advertising Age*, 27 Dec.

Brandon, J.A., Jones, W. and Ohman, M.D. (2019) 'Multidecadal increase in plastic particles in coastal ocean sediments', *Science Advances*, 5(9). doi:10.1126/sciadv. aax0587

Break Free From Plastic (2021) *Branded: Brand audit report 2021 – Vol IV.* Break Free From Plastic. Available at: www.breakfreefromplastic.org/brandaudit2021/

British Plastics Federation (2016) *The UK plastics industry: A strategic vision for growth.* British Plastics Federation. Available at: www.bpf.co.uk/plastics-strategy/default.aspx

British Plastics Federation (2022) *BPF: About the British plastics industry.* Available at: www.bpf.co.uk/industry/default.aspx#:~:text=It%20has%20an%20 annual%20sales,material%20processors%20and%20machinery%20manufacture

Carrington, D. (2022) 'Chemical pollution has passed safe limit for humanity, say scientists', *The Guardian*, 18 Jan. Available at: www.theguardian.com/environment/2022/ jan/18/chemical-pollution-has-passed-safe-limit-for-humanity-say-scientists

Charles, D., Kimman, L. and Saran, N. (2021) *The plastic waste makers index.* Minderoo Foundation. Available at: www.minderoo.org/plastic-waste-makers-index/

CNN Money (2021a) *Fortune 500: 1955 archive full list*. Available at: https://money.cnn.com/magazines/fortune/fortune500_archive/full/1955/1.html

CNN Money (2021b) *Fortune 500: 1965 archive full list*. Available from: https://money.cnn.com/magazines/fortune/fortune500_archive/full/1965/

CNN Money (2021c) *Fortune 500: 1975 archive full list*. Available from: https://money.cnn.com/magazines/fortune/fortune500_archive/full/1975/

CNN Money (2021d) *Fortune 500: 1985 archive full list*. https://money.cnn.com/magazines/fortune/fortune500_archive/full/1985/

Coca-Cola (2018) *The Cola-Cola Company: Moving toward a circular economy*. Available at: www.coca-colacompany.com/news/moving-toward-a-circular-economy

Coca-Cola Company (2022) *Brands and products: The Coca-Cola Company*. Available at: www.coca-colacompany.com/brands

Corcoran, P.L., Moore, C.J. and Jazvac, K. (2014) 'An anthropogenic marker horizon in the future rock record', *GSA Today*, 24(6), pp. 4–8. doi:10.1130/GSAT-G198A.1

Crippa, M., Solazzo, E., Guizzardi, D., Monforti-Ferrario, F., Tubiello, F.N. and Leip, A. (2021) 'Food systems are responsible for a third of global anthropogenic GHG emissions', *Nature Food*, 2, pp. 198–209. doi:10.1038/s43016-021-00225-9

Dauvergne, P. (2008) *The shadows of consumption: Consequences for the global environment*. Cambridge, MA: MIT Press.

Dickinson, T. (2020) 'Planet plastic: How big oil and big soda kept a global environmental calamity a secret for decades', *Rolling Stone*, 3 Mar. Available at: www.rollingstone.com/culture/culture-features/plastic-problem-recycling-myth-big-oil-950957/

Flanagan, R. (2020) 'What is and is not included in Canada's ban on single-use plastics', *CTV New*, 7 Oct. Available at: www.ctvnews.ca/climate-and-environment/what-is-and-is-not-included-in-canada-s-ban-on-single-use-plastics-1.5136387

Fortune (1996) 'Fortune's Global 500: The world's largest corporations', *Fortune*, 5 Aug.

Fortune (2021) 'G500: World's largest companies: The list', *Fortune*, Aug/Sept.

Foster, J.B. and Clark, B. (2020) *The robbery of nature: Capitalism and the ecological rift*. New York: Monthly Review Press.

Foster, J.B, Clark, B. and York, R. (2010) *The ecological rift: Capitalism's war on the Earth*. New York: Monthly Review Press.

Fraser, N. (2021) 'Climates of capital: For a trans-environmental eco-socialism', *New Left Review'*, 127(Jan/Feb), pp. 94–127.

Frito-Lay (2022) *Brands: FritoLay*. Available at: www.fritolay.com/brands

Geyer, R., Jambeck, J.R. and Law, K.L. (2017) 'Production, use, and fate of all plastics ever made', *Science Advances*, 3(7). doi:10.1126/sciadv.1700782

Goldman, R. and Papson, S. (1996) *Sign wars: The cluttered landscape of advertising*. New York: Guilford Press.

Goldsberry, C. (2021) 'Report blasts "false" corporate solutions to plastic pollution', *Plastics Today*, 28 June. Available at: www.plasticstoday.com/sustainability/report-blasts-false-corporate-solutions-to-plastic-pollution

Greenpeace (2018) *A crisis of convenience: The corporations behind the plastic pollution pandemic*. Greenpeace International. Available at: www.greenpeace. org/international/publication/19007/a-crisis-of-convenience-the-corporations-behind-the-plastics-pollution-pandemic/

Harvey, D. (2015) *Seventeen contradictions and the end of capitalism*. London: Profile Books.

Harvey, D. (2018) *Marx, capital, and the madness of economic reason*. New York: Oxford University Press.

Hawkins, G. (2013) 'Made to be wasted: PET and topologies of disposability', in Gabrys, J., Hawkins, G. and Michael, M. (eds.) *Accumulation: The material politics of plastic*. London: Routledge, pp. 49–67.

Hawkins, G., Potter, E. and Race, K. (2015) *Plastic water: The social and material life of bottled water*. Cambridge: MIT Press.

Heede, R. (2014) 'Tracing anthropogenic carbon dioxide and methane emissions to fossil fuel and cement producers, 1854–2010', *Climatic Change*, 122(1), pp. 229–241. doi:10.1007/s10584-013-0986-y

Johnson, B. (2020) 'World's largest advertisers 2020: Prime time: Amazon vaults into top spot, displacing Procter & Gamble', *Advertising Age*, 7 Dec.

Johnson, B. (2021a) 'They're back: Ad Age world's largest advertisers are spending again: The top 100 advertisers cut spending by 7.1% in 2020 amid the global pandemic. But ad spending is rising now as marketers rebuild and rebound', *Advertising Age*, 13 Dec.

Johnson, B. (2021b) 'The big list: Top marketers, brands and agencies. Ad Age Datacenter's definitive rankings of the biggest U.S. and global advertisers and agencies', *AdAge*, 16 Dec. Available at: https://adage.com/article/datacenter/big-list-top-marketers-brands-and-agencies/2383356

Lakhani, N., Uteuova, A. and Chang, A. (2021) 'Revealed: The true extent of America's food monopolies, and who pays the price', *The Guardian*, 14 July. Available at: www.theguardian.com/environment/ng-interactive/2021/jul/14/food-monopoly-meals-profits-data-investigation

Leiss, W., Kline, S., Jhally, S., Botterill, J. and Asquith, K. (2018) *Social communication in advertising*. London: Routledge.

Lewis, J. (2013) *Beyond consumer capitalism: Media and the limits to imagination*. Cambridge: Polity.

Lindeman, T. (2022) 'Canada lays out rules banning bags, straws and other single-use plastics', *The Guardian*, 20 June. Available at: www.theguardian.com/environment/2022/jun/20/canada-ban-single-use-plastics

Mah, A. (2022) *Plastic unlimited: How corporations are fuelling the ecological crisis and what we can do about it*. Cambridge: Polity.

Marchand, R. (1985) *Advertising the American dream: Making way for modernity 1920–1940*. Berkeley: University of California Press.

Marriott, J. and Minio-Paluello, M. (2013) 'Where does this stuff come from? Oil, plastic and the distribution of violence', in Gabrys, J., Hawkins, G. and Michael, M. (eds.) *Accumulation: The material politics of plastic*. London: Routledge, pp. 171–183.

Marx, K. (1990 [1867]) *Capital: A critique of political economy – Volume 1*. Translated by B. Fowkes. London: Penguin.

Moor, L. (2007) *The rise of brands*. Oxford: Berg.

Moore, J.M. (2015) *Capitalism in the web of life: Ecology and the accumulation of capital*. London: Verso.

Moore, J.M. (2016) 'The rise of cheap nature', in Moore, J.M. (ed.) *Anthropocene or Capitalocene? Nature, history, and the crisis of capitalism*. Oakland: PM Press, pp. 78–115.

Murdock, G. and Brevini, B. (2017) 'Carbon, capitalism, communication', in Brevini, B. and Murdock, G. (eds.) *Carbon capitalism and communication: Confronting climate crisis*. Cham: Palgrave Macmillan, pp. 1–20.

Nestlé (2022) *Our brands A-Z: Nestlé Global*. Available at: www.nestle.com/brands/brandssearchlist

Ocean Conservancy (2021) *We clean on: 2021 report*. Available at: https://oceanconservancy.org/trash-free-seas/international-coastal-cleanup/annual-data-release/

Packard, V. (1961) *The waste makers*. London: Longmans.

Papson, S.D. (1992) 'Green marketing and the commodity self', *Humanity & Society*, 16(3), pp. 390–413. doi:10.1177/016059769201600308

PepsiCo (2022) *PepsiCo: Product information*. Available at: www.pepsico.com/brands/product-information

Persson, L., Carney Almroth, B.M., Collins, C.D., Cornell, S., de Wit, C.A., Diamond, M.L., Fantke, P., Hassellöv, M., MacLeod, M., Ryberg, M.W., Søgaard Jørgensen, P., Villarrubia-Gómez, P., Wang, Z. and Hauschild, M.Z. (2022) 'Outside the safe operating space of the planetary boundary for novel entities', *Environmental Science & Technology*, 56(3), pp. 1510–1521. doi:10.1021/acs.est.1c04158

Prendergast, G. and Pitt, L. (1996) 'Packaging, marketing, logistics and the environment: Are there trade-offs?', *International Journal of Physical Distribution & Logistics Management*, 26(6). doi:10.1108/09600039610125206

Procter & Gamble (2022) *Brands: Procter and Gamble*. Available at: https://us.pg.com/brands/

Rosa, H. (2013) *Social acceleration: A new theory of modernity*. Translated by J. Trejo-Mathys. New York: Columbia University Press.

Rosa, H. (2017) 'Dynamic stabilization, the Triple A. approach to the good life, and the resonance conception', *Questions de Communication*, 31, pp. 437–456. doi:10.4000/questionsdecommunication.11228n

Rosa, H., Dörre, K. and Lessenich, S. (2017) 'Appropriation, activation and acceleration: The escalatory logics of capitalist modernity and the crises of dynamic stabilization', *Theory, Culture & Society*, 34(1), pp. 53–73. doi:10.1177/0263276416657600

Shove, E. (2003a) *Comfort, cleanliness and convenience: The social organization of normality*. Oxford: Berg.

Shove, E. (2003b) 'Converging conventions of comfort, cleanliness and convenience', *Journal of Consumer Policy*, 26(4), pp. 395–418. doi:10.1023/A:1026362829781

Simms, C. and Trott, P. (2010) 'Packaging development: A conceptual framework for identifying new product opportunities', *Marketing Theory*, 10(4), pp. 397–415. doi:10.1177/1470593110382826

Sinclair, J. (2012) *Advertising, the media and globalisation: A world in motion.* London: Routledge.

Sinclair, J. (2020) 'Cracking under pressure: Current trends in the global advertising industry', *Media International Australia*, 174(1), pp. 3–16. doi:10.1177/1329878X19873979

Steffen, W., Broadgate, W., Deutsch, L., Gaffney, O. and Ludwig, C. (2015) 'The trajectory of the Anthropocene: The great acceleration', *The Anthropocene Review*, 2(1), pp. 81–98. doi:10.1177/2053019614564785

Steffen, W., Grinevald, J., Crutzen, P. and McNeill, J. (2011) 'The Anthropocene: Conceptual and historical perspectives', *Philosophical Transactions of the Royal Society A*, 369(1938), pp. 842–867. doi:10.1098/rsta.2010.0327

Strasser, S. (1989) *Satisfaction guaranteed: The making of the American mass market.* New York: Pantheon Books.

Strasser, S. (2019) 'Never gonna give you up: how plastic seduced America', *The Guardian*, 21 June. Available at: www.theguardian.com/us-news/2019/jun/21/history-of-america-love-affair-with-plastic

Tedlow, R.S. (1990) *New and improved: The story of mass marketing in America.* Oxford: Heinemann Professional.

Trentmann, F. (2017) *Empire of things: How we became a world of consumers, from the fifteenth century to the twenty-first.* London: Penguin Books.

Twede, D. (2012) 'The birth of modern packaging: Cartons, cans and bottles', *Journal of Historical Research in Marketing*, 4(2), pp. 245–272. doi:10.1108/17557501211224449

Twede, D. (2016) 'History of packaging', in Jones, D.G.B. and Tadajewski, M. (eds.) *The Routledge companion to marketing history.* London: Routledge, pp. 115–129.

Unilever (2022a) *Brands: Unilever.* Available from: www.unilever.com/brands/all-brands/

Unilever (2022b) *Rethinking plastic packaging: Unilever.* Available from: www.unilever.com/planet-and-society/waste-free-world/rethinking-plastic-packaging/

Unruh, G.C. (2000) 'Understanding carbon lock-in', *Energy Policy*, 28(12), pp. 817–830. doi:10.1016/S0301-4215(00)00070-7

Weber, M. (2021) 'The cult of convenience: Marketing and food in postwar America, *Enterprise & Society*, 22(3), pp. 605–634. doi:10.1017/eso.2020.7

Wernick, A. (1991) *Promotional culture: Advertising, ideology, and symbolic expression.* London: SAGE.

Willis, S. (1991) *A primer for everyday life.* London: Routledge.

Wu, T. (2018) 'The tyranny of convenience', *The New York Times*, 16 Feb.

2 E-commerce and Acceleration

Expediting Circulation, Normalising Fossil-Fuelled Convenience

The e-commerce boom, further spurred by the COVID-19 pandemic lock-downs of 2020–21, has radically transformed how many of us browse and buy things, how we are marketed to, and how commodities move from the factory floor to domestic spaces. E-commerce giants such as Amazon and a range of supporting logistics and courier companies administer and execute financial transactions and the physical delivery of goods, playing an increasing role in the emerging configuration of corporate retail power. And old retail powerhouses have not simply gone away. Already powerful entities are taking advantage of new conditions. Walmart ranked number one in *Fortune's* 2021 Global 500 ranking (for 2020 revenues) (Fortune 2021), fortified by its status as number two in U.S. e-commerce behind Amazon (Wahba 2021). Walmart, a combination of bricks-and-mortar and online retailer, beat out Big Oil and Big Tech in terms of revenue generation. Together, what I will call Big Retail and Big Logistics support the expedited circulation of commodities and accumulation of capital in contemporary consumer societies. Rather than focusing on digital shopfronts and market-places, my examination of e-commerce addresses business logistics – the distribution and transportation arrangements that support accelerated commodity exchange. It is not enough for commodities to be made desirable with symbolic meanings; capitalists also need convenient retail and delivery systems to ensure that commodities are smoothly and speedily ushered from producer to consumer. By highlighting the physical movement of products to people, I underline spatial, temporal, and material dimensions, and hence the resource intensiveness, of e-commerce.

This chapter considers how the coordination of digital technologies, logistical systems, and material infrastructures characteristic of e-commerce works to supersede existing spatial and temporal boundaries, accelerating the circuit of production, exchange, distribution, and consumption. E-commerce and logistics giants are agents of convenience and speed. As we saw in Chapter 1, capitalist acceleration in the nineteenth and twentieth

DOI: 10.4324/9781003001621-3

centuries was powered by fossil fuel consumption. With online shopping, expedited consumption necessitates intensified reliance on energy-hungry infrastructures and systems of production and distribution, and hence, fossil fuel dependence. Industrial dynamics and features expressed through e-commerce and logistics, and the capitalist growth imperative underpinning them, throw up problems for the material world. Drawing on David Harvey's (2018: 1–23) reading of Marx and capital as 'value in motion', I will conceptualise digital market exchange and fossil fuelled delivery as means for accelerating the realisation of value and circulation of commodities. Building on an examination of the case of Amazon, I will reflect on how the normalisation of convenience (e.g., ease of shopping experience) and speed (e.g., one- and two-day delivery) entrench resource-intensive overconsumption. Fostering a more balanced and sustainable relationship to the planet calls for a move in the opposite direction.

E-commerce, Big Retail, and Big Logistics

While the online shopping sites that sell goods and administer transactions may be the face of e-commerce, just as important are the logistics and transportation companies that enable the delivery of those goods. E-commerce is defined as:

> the sale or purchase of goods or services, conducted over computer networks by methods specifically designed for the purpose of receiving or placing orders. . . . Accordingly, whether a commercial transaction qualifies as e-commerce is determined by the ordering method rather than the characteristics of the product purchased, the parties involved, the mode of payment or the delivery channel.
>
> (OECD 2019: 14)

A complex system of inventory management, physical distribution, and tracking is needed to make good on a process set in motion by a digital sale: business logistics and supply chain management enable acquisition of physical commodities. Before considering some of the powerful companies involved in the online shopping sector, I will briefly highlight key dynamics shaping contemporary business logistics more generally, building on the explanation of logistics provided in the introductory chapter.

Logistics involves moving products but encompasses much more than freight transport. Logistics management 'seeks to optimize the flows of materials and supplies through the organization and its operations to the customer' (Christopher 2010: 3). This 'planning process' and 'information-based activity' (Christopher 2010: 3) is supported by advanced information technology (IT), including 'sophisticated supply chain management soft-ware, Internet-based logistics systems, point-of-sale scanners, RFID [radio-frequency identification] tags, satellite tracking, and electronic transfer of order and payment data' (Armstrong, Kotler, and Opresnik 2016: 348). IT systems used in the administration of digital sales and ordering systems keep goods moving smoothly and quickly and reduce reliance on invento-ries and associated costs:

> as an item is sold and scanned in a shop, the data are used to inform replenishment and re-ordering systems and thus react quickly to demand. Sharing such data with key suppliers further integrates pro-duction with the supply function.
>
> (Fernie, Sparks, and McKinnon 2010: 895)

The moments of exchange, distribution, and consumption exist in a tight feedback loop with the moment of production.

Powerful retailers can 'control, organise and manage the supply chain from production to consumption' (Fernie, Sparks, and McKinnon 2010: 895), and exert more pressure on logistics companies in efforts to maximise efficiencies and minimise costs (Newsome 2015: 30–32). With e-commerce, Big Retail and Big Logistics both have been beneficiaries of the reconfigu-ration of shopping around digital sale and individualised physical delivery.

E-commerce has been a growth industry in economically volatile times. As reported in *Forbes*, global e-commerce sales for 2021 were forecast to be $4.2 trillion, after achieving year-over-year first quarter growth of 38% (Verdon 2021). eMarketer (2021b) predicted a higher global figure – near $5 trillion for 2021 – following up a 27.6% rate of growth in 2020. According to eMarketer (2020), in the United States in 2020, '[o]nline shopping [was] so strong that it [would] more than offset the 3.2% decline in brick-and-mortar spending [in 2020]'. Some of this industry growth was a product of the historical specificity of the COVID-19 pandemic – for instance, the spending of stimulus payments in the United States, safety-driven prefer-ence for buying items online, and increased demand for home-based enter-tainment. In the United States, 'consumers spent $1.7 trillion online, $609 billion more than the two preceding years, between March 2020 to February 2022' (Adobe 2022) (though rising prices connected to inflation contribute to this increase). The rise of online grocery sales, 'a category with minimal discounting', reveals how 'speed and convenience are becoming just as impor-tant as cost savings' in e-commerce trade, according to an Adobe marketing expert (Brown in Adobe 2022). E-commerce growth is projected to continue, with industry analysts anticipating that new shopping behaviours shaped by current circumstances will endure, with consumers who had never made an online purchase before the pandemic continuing to shop online (Verdon 2021; eMarketer 2020).

E-commerce is prevalent in the world's largest economies, and in some of the most quickly consumed, fashion-oriented, and disposable product categories. China leads in global e-commerce sales, followed by the United States, the United Kingdom, Japan, and South Korea (Edmondson 2021). Apparel, footwear, and food and drink reportedly will be central drivers of e-commerce growth during the 2020–2025 period (Evans 2021). Investment specialist Jane Edmondson (2021) cites 'competitive pricing, shopping convenience, greater product selection and rapid delivery options' as factors that 'have solidified online commerce as a disruptive technology that is here to stay'.

One cannot safely predict what will happen with e-commerce, as industry forecasts are necessarily speculative. However, as consumers get used to the convenience and speed of online shopping, formerly novel perks may become normalised as standard expectations. The demands of online retailers and consumers, together, are shaping high standards of service delivery by which logistics and courier companies must abide, as is strikingly captured in an advertisement by DHL, a giant in package delivery and logistics, quoted below:

> It hasn't been such a long time since people first started shopping online, and it was easy to keep up with the clicks. Little by little by little by more the clicks added up to be more than the stores, and e-tailers worked harder than ever before, because they had to keep up with the clicks ... Monday through Sunday, office and home, 24 hours in every zone, and still they kept up with the clicks. ... All of a sudden there were so many 'must do's' if you want to keep selling – like delivery *there*, no *there*, or *there*. [Switches to client/customer] 'And make it tomorrow – first thing'.
>
> (DHL 2021)

Convenience is reimagined as an absence of waiting, with customers and retailers setting the fast pace to which logistics and transportation companies must respond. Indeed, over a decade ago, retailing and logistics scholars John Fernie, Leigh Sparks, and Alan C. McKinnon (2010: 906) remarked on the challenges tied to 'last mile' delivery (i.e., via courier to the customer's doorstep): 'Most customers would like deliveries to be made urgently at a precise time with 100 per cent reliability'. Such perceived urgency sets e-commerce apart from its non-digital precursor, catalogue mail order-based home shopping.

The Big Retail power brokers profiting from e-commerce have an obvious interest in continuing to encourage this shopping practice. Just ten e-commerce companies reportedly captured the majority of U.S. e-commerce sales in 2020: Amazon (39%); Walmart[1] (5.8%); eBay (4.9%); Apple (3.5%); The Home Depot (2.1%); Best Buy (2.0%); Target (1.7%); Wayfair (1.5%); The Kroger Co. (1.4%); and Costco Wholesale (1.4%) (eMarketer 2020). Here we see a mixture of online and more conventional retailers and marketplaces, with Amazon accounting for more online trade/sales than the others combined in the U.S. The largest players cashed in on the pandemic.

Big Logistics, including the third-party logistics (3PL) sector (specialists to whom sellers outsource distribution and supply chain management functions), likewise have reaped the rewards of changing patterns of consumption. For instance, UPS, the U.S. Postal Service, Deutsche Post DHL, and FedEx ranked 89th, 123rd, 113th, and 135th on Fortune's 2021 Global 500 (based on 2020 revenues), respectively, and China Post Group ranked 74th (Fortune 2021). By comparison, Facebook ranked 86th, Walt Disney ranked 145th, and Procter & Gamble ranked 128th (Fortune 2021). Moving and monitoring commodities is big business, and e-commerce is driving third-party logistics growth. The 2021 first quarter results for DHL were the logistics company's 'best-ever' (Dempsey 2021). UPS and FedEx also experienced increases attributable to online shopping overall (Dempsey 2021). The strain on existing logistical capacity produced by sector growth drove further expansion of capacity. For instance, DHL 'increas[ed] investment in its network to meet the bigger demand for its services' (Dempsey 2021), and UPS and FedEx expanded their workforces (Bentz 2021).

While sector growth and expansion may slow, plateau, or even decline, proving to be unique to an historical moment shaped by pandemic conditions, online shopping will nevertheless remain a now familiar option experienced and expected by many across the globe. During a period of extraordinary economic disruption, e-commerce helped ensure that commodities made their way to and from manufacturers, retailers, and consumers, safeguarding the continued productivity of the capitalist system and dominant corporations. During a moment that presented an opportunity to rethink fossil fuel-dependent commercial systems and infrastructures, consumer societies instead doubled down on this reliance.

Capitalist Consumption and 'Value in Motion'

To grasp the larger political economic forces and pressures governing the incessant pursuit of growth exemplified by the e-commerce and logistics sectors, we need to understand how capital circulates and accumulates under capitalism. Adopting a Marxian lens, we can understand capital as 'value in motion' (Harvey 2018), a process to which contemporary logistics companies and their workers contribute (Newsome 2015).[2] Karl Marx (1992

[1885]: 185) characterises capital as 'a movement, a circulatory process through different stages'. According to Harvey's (2018: 4–23) explanation, this process involves the transformation of money capital into commodities and surplus value in production; the realisation of value in money form in exchange; the distribution of this money to various capitalists (industrialists, merchants, landlords, financiers) in profits and to workers in wages; and the rendering of some of the capitalists' money into money capital used to produce more commodities. '[T]he circulation of capital is inherently a logistical affair' that is characterised by acceleration: 're-organizations of space and time' have been brought about by technologies and systems involving 'jet transportation, container shipping, and digitization' (Manzerolle and Kjøsen 2012: 219). The ongoing cycle of reinvestment and capital accumulation is not seamless; the capitalist system is prone to crisis, and capitalists are tasked with the perpetual need to create new markets. The transformation of logistics that has transpired since the economic crises of the 1970s can be interpreted as a 'reorganization of capitalist circulatory systems', where 'increases in the speed, cost-efficiency, volume, reliability, and flexibility of commodity flows' helped keep capital accumulating and value circulating (Danyluk 2018: 635).

Technological advancements in logistics in terms of IT and transportation have supported the tightening feedback loop between production and consumption. In the capitalist system more generally, technology plays an important role in the ongoing cycle of production and reinvestment, with innovations serving 'to discipline and disempower labour, to raise the productivity of labour and to increase the efficiency and speed of turnover of capital in both production and circulation' (Harvey 2018: 120). '[T]echnology itself becomes a business,' Harvey points out: 'it produces a commodity – new technologies or organisational forms – that need to find and even create a new market' (Harvey 2018: 119). Likewise, transportation 'both produces a commodity – the movement of goods and people – and helps in the production and realization of value' (Pitts 2015: 209–210). Successful commodity exchange is contingent on transportation and distribution processes that bring about commodity acquisition.

To meet the high levels of customer service now widely expected (beyond e-commerce alone), such processes and systems must be flexible as well as fast. As geographer Martin Danyluk (2018: 638) explains:

> corporations' embrace of "agile" or "nimble" supply chains can be understood as a method of opening up new opportunities for the realization of value by dynamically reshaping the spatial pathways through which commodity capital circulates. Many of the new transportation,

warehousing, and retailing practices that make up the logistics revolution are underpinned by this logic.

Innovations in logistics intensify and quicken a longstanding tendency in capitalism 'to enhance the mobility of commodities' (Danyluk 2018: 639). Importantly, the fast movement of commodity cargo across international trade networks has not only been a product of new technologies and deft planning and information management. Supporting energy and transportation infrastructures have been created and fossil fuels consumed in order to power the rapid movement of goods across the globe.

The production of new commodities and realisation of value – and the accelerating circulation of capital – necessitate the consumption of energy, natural resources, and various materials. Thus, value in motion is also matter and energy in motion, or as Jason W. Moore (2015: 174) puts it, 'Capital is value-in-motion is value-in-nature'. Environmental impacts are incurred at each 'moment' in the ongoing circuit of commodity production, exchange, distribution, and consumption, not just 'final consumption' (e.g., consumer purchases of goods ranging from necessities to luxuries) (Harvey 2018: 13). Thus, investment and profit seeking by capitalists, which supports yet more production, is a key part of 'the environmental equation' (Foster, Clark, and York 2010: 382).

As we saw in the introductory chapter, Marx conceptualises 'productive consumption' as distinct from final consumption; it involves commodities consumed by capitalists in the process of producing more commodities (Harvey 2018: 12). Ecologically consequential consumption by capitalists encompasses the processes not only of manufacture but also distribution and retail, which make commodity exchange possible. Under capitalism, land is also treated as a commodity, and producing the infrastructures required for physical distribution of e-commerce sales requires considerable consumption of land. Amazon's distribution centres, for instance, occupy anywhere from '50,000 to more than 1.2 million square feet, depending on their function and their role in the specific setting of the regional network' (Hesse 2018: 408). Repurposing land in this way curtails alternative uses, such as preserving or creating natural carbon sinks such as forests and using land to support biodiversity.

As a retailer, marketplace, and distributor, not a manufacturer, corporate behemoth Amazon holds oligopoly power not over commodity production but instead commodity exchange and distribution. 'Merchant capitalists' assume considerable power due to their vital role in the realisation of exchange and in propelling value in motion, according to Harvey (2018: 35):

> The time taken to get the product to market and achieve a sale is lost time and time is money. For this reason industrial capitalists often prefer to

pass the commodity on immediately to merchants. The merchant capitalist organises sales in an efficient way and at a low cost (chronically exploiting labour power in the process). The creation of warehouses, department stores and delivery services (now increasingly online) produces economies of scale in marketing. . . . [I]ndustrial producers have a strong incentive to pass on their commodities to merchants at a discount of the full value prior to the moment of realisation.

As a merchant capitalist with interests also spread across an expansive, controlled system of warehouses, distribution centres, and delivery services, Amazon has played a crucial role in accelerating the circulation of commodities. As geographer Markus Hesse (2018: 411) observes:

Amazon.com is accelerating supply chain processes, in order to get even closer to the customer, by annihilating space and compressing time. It now adds a completely new layer of distribution to the system, situated in between large-scale fulfilment centres and the customer. Small-scale sorting and delivery centres allow keeping the promise of same-day service or even shorter delivery time frames.

In the next section, I will examine the case of Amazon, and the industry standards of convenience and speed it has set in the e-commerce and logistics sectors, in more detail. As the industry leader, Amazon has helped lubricate and accelerate the entire feedback loop between production and consumption.

Raising Standard Expectations: The Case of Amazon

Amazon brings into sharp focus both essential capitalist and specific e-commerce logics and contradictions. Amazon's divisions traverse e-commerce, logistics, streaming, and cloud computing, capitalising on business-to-business (B2B) and business-to-consumer (B2C) markets, and hence, final and productive consumption. Amazon is one of the 'big five' U.S. Tech Giants alongside Apple, Facebook, Google, and Microsoft, but its overwhelming dominance of the e-commerce market distinguishes it. Amazon sells its own product lines and provides online retail space and delivery services to third-party sellers:

More than half of the goods the company sells are through the Amazon Marketplace and Fulfillment by Amazon (FBA) programs, the former where sellers use the site to sell their goods but take care of delivery themselves, and the latter where sellers also pay Amazon to warehouse their goods and arrange their delivery.

(West 2022: 9)

The mass adoption of Amazon Web Services (AWS) – a subsidiary that sells subscription use of its cloud computing infrastructure and services to businesses, governments, and other organisations – has shored up and expanded Amazon's corporate reach and power (Brevini 2021a: 9–10, 14). Amazon is also remaking the logistics sector to suit its own needs. Through the many services Amazon licenses and collects rents on, Amazon supplies the means by which other companies can produce, exchange, and distribute commodities (Brevini 2021b: 3–4).

Amazon is a giant according to a number of measures. Amazon has operations and customers across the planet. The corporation dominates globally in terms of revenue, ranking ninth in *Fortune*'s 2020 edition of the Global 500 and securing the third spot on the 2021 list (Fortune 2021). Amazon is also a global leader in terms of advertising spend. In 2019, Amazon outspent consumer goods corporation Procter & Gamble on advertising and promotion, allocating $11 billion to these activities and ranking number one in the Ad Age World's Largest Advertisers report (Johnson 2020). 'Amazon has been among the biggest beneficiaries of the coronavirus pandemic, as crowd-averse shoppers rushed online', a *Fortune* contributor observes, with the number of Prime subscribers rising from 150 million at the beginning of 2020 to 200 million as of April 2021 (Day and Bloomberg 2021). These subscribers can use Amazon's varied services:

> Prime members . . . order clothing, staples, and electronics from the e-commerce giant, watch original Prime Video movies and TV shows, and listen to music on Amazon's streaming media channels. Even consumers who don't actively use Amazon's website to shop spend much of their digital lives using services like Netflix that run on Amazon's ubiquitous AWS servers. Now the pandemic has accelerated these trends more than anyone could have imagined.
>
> (Dumaine 2020: 89)

Customer service standards set by this corporate juggernaut influence online shopping, logistics, media, and cloud computing.

Amazon sells a variety of commodities (goods and services) in B2B markets. Amazon functions as an advertising platform – a means of promotion for other companies. According to eMarketer (2021a), as of 2020, Amazon ranked third in U.S. digital advertising behind Google and Facebook, in what the market research company characterises as a triopoly. Through advertising sales including search, Sponsored Products, and Sponsored Brands, Amazon realised advertising revenues of $15.73 billion in 2020 (eMarketer 2021a).

Data storage and computing power are commodities on which both business customers and end consumers rely, and Amazon is a colossal player in the data centre and cloud computing sector. Amazon led a massive shift from the *private cloud*, which 'is customized and deployed for a single organization', to the *public cloud*, which 'is typically provided by large cloud-services businesses . . . and offers software, platforms, and infrastructure to the general public or to an industry association' (Mosco 2014: 40–41). Corporate AWS customers range from financial services companies HSBC and Capital One to streaming services Netflix and Peacock to online food delivery company Just Eat (AWS 2021).

Turning to B2C markets, Amazon operates a media streaming service: Prime Video. This service features Amazon's own commissioned content but also film, TV, and sports programming owned by other media companies. As Amazon's Jeff Bezos indicated in an April 2021 press release, 'over 175 million Prime members . . . streamed shows and movies in the past year, and streaming hours are up more than 70%' from the previous year (in Amazon 2021e). Media content is used as a bonus provided to shoppers who elect to pay for Amazon Prime, a subscription programme that offers members fast, free shipping – and vice versa. Media streaming promotes consumer demand for Amazon's e-commerce interests, and online shopping promotes demand for Amazon's streaming service.

Amazon dominates the global e-commerce market, where it serves as a market maker and means of market exchange for other manufacturers and brands as well as selling to end consumers. Amazon is a gatekeeper for online sales, with third-party sellers 'account[ing] for 58% of Amazon sales' and third-party sales 'growing at 52% a year (compared to 25% for first-party sales by Amazon)' (Amazon 2021c). A *Forbes* contributor characterises Amazon as ostensibly 'the default shopping channel for many consumers. If brands aren't on Amazon, they could be losing out to competitors who are' (McAfee 2021). Amazon collects subscription fees, selling fees, and Fulfillment by Amazon (FBA) fees from third parties (Amazon 2021c). The FBA service allows for use of Amazon's fulfillment centres and associated inventory management, packing, and shipping services by third-party sellers (Amazon 2021d). Amazon collects fees for inventory storage, long-term storage, preparing and shipping orders, returns, and so on, and offers third-party sellers eligibility for the free two-day shipping policy that Amazon Prime subscribers expect (Amazon 2021d). And Amazon can ship faster than that, offering same-day and even two-hour delivery times:

> The reality and materiality of all these goods moving through space arguably has been lost as we experience them only via the metric of time.
>
> (West 2022: 61)

Alternative metrics, such as carbon dioxide emissions, are profoundly undervalued.

E-commerce market strength is bolstered by Amazon's expansive and dynamic logistical system, which enables the company to 'reduc[e] friction in purchase and time from click to ship to delivery' (West 2022: 54). Amazon's logistics system encompasses delivery vehicles, warehouses, and predictive analytics systems that claim to 'understand and predict buyer behavior and demand so that it can put goods in the hands of consumers faster than ever' (McAfee 2021). Amazon operates 'fulfilment centres' in the United States, Canada, Mexico, Brazil, the United Kingdom, Ireland, Germany, France, Spain, Poland, Slovakia, Sweden, Japan, India, Australia, the United Arab Emirates, Egypt, Kuwait, and Saudi Arabia (MWPVL 2021). These massive warehouses and distribution centres are highly technologised, deploying robotics, scanners, and computer systems in concert with warehouse workers to move commodities rapidly from seller to buyer (West 2022: 69–72).

This infrastructure used to expedite the speed of shopping takes up considerable space, as noted above. For instance, in the United States, Amazon's sprawling system of warehouses and distribution centres spans '173 million square feet and another 1,100 globally . . . cover 262 million square feet' (Davis 2020). Further changes to residential landscapes are on the horizon, with Amazon planning 'to open 1,000 small delivery hubs in cities and suburbs all over the U.S.', reports *Bloomberg.com*, 'bring[ing] products closer to customers, making shopping online about as fast as a quick run to the store' (Soper 2020). Optimisation of shopping for speed has considerable land use implications.

The process of expediting delivery is also reliant on high-speed transportation, and Amazon 'is amping up their presence in the logistics space with their own fleet of planes, trailers and delivery trucks' (Bentz 2021). In addition to operating its own fleet, Amazon enters into delivery agreements with other couriers. It also 'lease[s] Amazon-branded vans' to individuals who wish to run their own delivery businesses (Amazon 2021b). According to research conducted by the Bank of America, 'Internationally, Amazon handled 48% of its own deliveries – 1.2 billion of the 2.5 billion packages it shipped outside of the U.S. in 2019' (Davis 2020). In the US delivery services sector, Amazon trails only FedEx, UPS, and the United States Postal Service, and while its focus is on delivering Amazon sales, it is poised to expand this delivery service to other online retailers (Davis 2020). As one logistics consulting executive observed, 'In just a few years, Amazon has built its own UPS' (in Soper 2020).

A similar dynamic can be identified in Amazon's approach to producing and selling infrastructure in its cloud computing and logistics businesses. As a *Forbes* contributor observed:

AWS [Amazon Web Services] was initially developed to solve an internal constraint at Amazon, and once the infrastructure was built, Amazon could sell incremental excess capacity on third parties like private companies and government. Similarly, Amazon developed its logistics systems to solve its own problems, and can resell capacity to other parties.

(Masters 2020)

The dynamic at work – building internal capacity until an excess is produced – generates new products and markets for external buyers: final and productive consumption, to use the terminology introduced above.

Amazon has helped reshape the rhythm and literal landscape of consumer society. As geographer Hesse points out:

By using data centres to direct the information flow, and deploying fulfilment centres to navigate the consolidation and physical movement of consignments, *Amazon.com*'s network creates an entirely new geography of distribution.

(Hesse 2018: 407)

A 'logistical fix' (Danyluk 2018) has helped keep value in motion, accelerating the circulation of commodities and capital. Amazon brings together predictive technologies and logistics to spur sales, mediate digital transactions, and expedite the delivery of digital and physical commodities. Low prices, a convenient shopping experience, and 'a one-click "Buy Now" option and free delivery on Prime goods' have enabled Amazon to eliminate 'many elements of friction from the buying experience' (McAfee 2021). Amazon's success at delivering such a high level of customer service has created enormous pressure for other companies to follow suit. Rather than trying to compete with Amazon, many businesses instead purchase its e-commerce and logistics services.

From the perspective of competing businesses, it appears that increased customer convenience comes at a cost. Through its ability to expand operations via aggressive capital investment in business infrastructure, Amazon has built capacity to collect rent on a wide range of Amazon-owned services, as indicated above. Having expanded its logistics division, further uniting e-commerce and logistics, Amazon has extended its reach across the retail and distribution system, and now exercises considerable control over access to the means of making a profit and making a living. For example, Amazon can impose restrictions limiting the trade and decision-making of other e-commerce and logistics businesses. In an exercise of control, Amazon temporarily prohibited third-party suppliers on Amazon from using FedEx due to what Amazon deemed 'slipping performance' (Masters 2020) – a ban that was later lifted, affecting FedEx share values (Palmer 2020). The broader takeaway is that Amazon sets the pace and terms under which many other companies, including third-party logistics firms, must perform. As a *Bloomberg.com* e-commerce reporter observes (Soper 2020):

> Beyond Amazon's retail rivals, the mass opening of small, quick-delivery warehouses poses a significant threat to United Parcel Service Inc. and the U.S. Postal Service. Being fastest in the online delivery race is so critical to Amazon's business that it doesn't trust the job to anyone else and is pulling back from these long-time delivery partners.

While proper treatment of the topic lies outside the scope of this chapter, it warrants noting that the new configuration of retail power promises to erode public sector provision via post office services.

E-commerce and the logistical systems and infrastructures underpinning it have expanded the time and space of delivery, driving various forms of consumption. As noted in trade journal *Logistics Management*, 'global e-commerce behemoths have moved into logistics in a big way, . . . significantly raising the bar in customer expectations' (Banks and Hajibashi 2021). Given its size, scope, and influence, I have explored the case of Amazon as a way into understanding how e-commerce expedites value in motion. Amazon intervenes across supply chain processes, crucially, setting the pace at which other companies must deliver orders by making two-day, one-day, and even same-day delivery options standard. In so doing, Amazon has created new expectations of consumer service that, once normalised, would be challenging to reverse.

Interestingly, Amazon attached a copy of its 1997 letter to shareholders to its 2020 Annual Report, in which Jeff Bezos set out a long-term vision for the company: 'Today, online commerce saves customers money and precious time. Tomorrow, through personalization, online commerce will accelerate the very process of discovery' (Amazon 2021a). Bezos advised his shareholders: 'Market leadership can translate directly to higher revenue, higher profitability, greater capital velocity, and correspondingly stronger returns on invested capital' (Amazon 2021a). In the years since, Amazon has come to dominate B2B and B2C markets, transforming consumption by businesses and consumers.

Today, Amazon is a fulcrum in e-commerce, exercising market leadership across three distinct but interwoven oligopolies – Big Retail and Big Logistics in addition to Big Tech – and using its arsenal of assets to speed up the cycle of production, exchange, distribution, and consumption. In an age of ecological crisis, a key area of concern is the environmental fallout of the aggressive consumption of energy and natural resources that props up this entire system. Hesse (2018: 413) warns:

> the idea that the marriage of big data and sophisticated distribution creates a new super-monopoly in private hands bears a fundamental concern for open societies: the machine is now operated, controlled and valorized by a private enterprise, fit for profit and subject to the relentless exploitation of resources, human and non-human, urban and non-urban.

There is a contradiction between economic growth and environmental sustainability, and big business, including corporate monopolists like Amazon, pursues the former first and foremost.

Paying for Convenience: The Ecological Implications of High-Speed Shopping

Under capitalism, the circulation and accumulation of capital must grow and expand. As we see with e-commerce, digital technologies and business logistics are deployed to minimise friction and accelerate the process, arguably encouraging consumers to 'fetishize time . . . and mystify space along with the infrastructure and labor that it takes to "annihilate" it' (West 2022: 61). The systems and infrastructures supporting commodity exchange and distribution come at an environmental cost. According to one marketing textbook:

> more than almost any other marketing function, logistics affects the environment and a firm's environmental sustainability efforts. Transportation, warehousing, packaging, and other logistics functions are typically the biggest supply chain contributors to the company's environmental footprint.
>
> (Armstrong, Kotler, and Opresnik 2016: 349)

Opinion is divided on whether the popularisation of e-commerce will improve or aggravate this environmental footprint. More optimistic research on logistics, transportation, and cleaner production champions potential environmental gains from a projected decline in passenger travel to and from shopping districts, whereas soberer analyses highlight the carbon dioxide emissions and energy consumption generated by personalised freight transport, product returns, unsuccessful delivery attempts, and the excessive packaging associated with e-commerce (see overview in Pålsson, Pettersson, and Hiselius 2017).

While the situation is undoubtedly complex, by focusing on the larger capitalist pressure to grow, we can identify the prevailing direction of travel in contemporary consumer societies: intensified circulation and accumulation of commodities. By considering productive consumption by capitalists as well as final consumption by consumers, we can see how potential reductions in emissions in some areas (for instance, if online shoppers were to drive less) may be cancelled out by increases in the production, distribution, *and* consumption of commodities overall.[3] Such increases in e-commerce trade are the goal around which high levels of customer service and supporting technologies are mobilised. The need for more growth sets in motion the continual production and circulation of new products and markets, which, in turn, contributes to escalating consumer expectations of convenience encouraged by easy access to various products and markets. As logistics specialists Fernie, Sparks, and McKinnon (2010: 903) remarked over a decade ago, online shopping is 'substantially increasing the volume of goods that must be handled, creating the need for new DCs [distribution centres] and larger vehicle fleets', and is characterised by 'high logistical expectations' from customers in terms of the speed, reliability, and convenience of delivery – trends that have subsequently deepened and become normalised.

The resource intensiveness of the broader system configured for the provision of more, faster, and easier – *en masse* – bodes ill in the context of climate change, growing resource scarcity, mass extinctions, and already dangerous levels of pollution. Continued investment in the expansion of infrastructures of commodity consumption produces yet more consumer expectations. The assumption that one practice (driving to and from retail districts) will straightforwardly be substituted for another (goods being driven to us in residential neighbourhoods) points to a broader problematic tendency, especially in the private sector, to assume that environmental sustainability can be achieved through substitutions, new technologies, and efficiency gains. Genuine alternatives, such as creating better infrastructure and improving urban planning to support walking, cycling, and the use of public transport, are not the focus. After all, such approaches present more challenging paths to capital accumulation. Importantly, despite the efficiency gains delivered by generation after generation of new technologies, ecological crises have only worsened.

By taking a longer historical perspective, we can better understand why. As Hubert Buch-Hansen (2019: 44–45; emphasis added) explains, the 'Jevons paradox', named after nineteenth-century economist William Stanley Jevons, extends Jevons's observation that:

> although every new steam engine that was produced was more efficient in terms of coal use than the already existing steam engines, the production of an increasing number of steam engines meant that coal

usage increased instead of declining (Jevons 1865). In other words, *as technology becomes more efficient, production tends to expand.*

Recall Harvey's (2018: 119) observation that once technology becomes a business, it must develop new products and markets in order to grow. Capitalist investment in 'artificial intelligence, blockchain, the Internet of Things and autonomous delivery devices like drones or robots' (OECD 2019: 9; Bentz 2021), for instance, provides an additional suite of commodities that support capital accumulation and the circulation of other commodities. As commodities, technologies also exist in the cycle of production, consumption, and replacement, with new technologies involving particularly high rates of obsolescence. The production and eventual replacement of 'greener' technologies likewise sets in motion a circuit of production, exchange, distribution, and consumption. We can observe that '[i]nnovations that improve efficiency and coordination, or accelerate turnover times in both production and circulation, yield more surplus values for capital' (Harvey 2018: 110). Hence, 'greater efficiency is almost always accompanied by more consumption, reducing or even canceling out gains' (Klein 2019: 86). As cars have become more fuel efficient, people have been able to afford to drive more often and farther. With respect to courier services, efficiency gains are counteracted by more frequent home deliveries (and returns and redeliveries). Technological fixes are undermined, at least in part, by the growth imperative.

More transformative approaches would focus on consuming less, walking more, and so on. Such shifts would require very different social conventions. Instead, in advanced capitalist economies and among the affluent, the normalisation of high-speed delivery and multiplying conveniences as default expectations, not novel perks, further engrains resource-intensive consumption. As Elizabeth Shove (2003: 170) observes, 'reliance on convenient solutions has the cumulative effect of redefining what people take for granted.' It becomes hard to turn back and make do with what is perceived as less.

Escalating expectations of customer service accompanying the popularisation of online shopping are producing *inefficiencies* in the approaches to package delivery in terms of load size and energy use/carbon consumption. As conceded in an e-commerce industry blog:

> Since there is a large emphasis on the importance of immediacy in business – especially in shipping offerings – businesses may have to send out freights that are only partially full. This will require additional trips and more transportation emissions.

(Collins 2021)

The rise of e-commerce is encouraging smaller shipment sizes, with parcel and 'less-than-truckload' (LTL) shipments becoming increasingly common (Bentz 2021). With direct delivery to consumers in residential areas, there are fewer packages per stop (less density), and 'networks are being pushed beyond capacity in many areas by the time it takes to complete deliveries on low-density routes' (Bentz 2021). Increased item returns (recall the popularity of clothing and shoes) compound the problem of personalised delivery, leading to yet more delivery trips (Collins 2021).

Furthermore, the advice to fly and drive less that many individuals are heeding in response to the climate crisis does not seem to extend to our commodity cargos. In addition to using smaller courier vehicles to complete the final leg of the delivery, air cargo and semi-trailer trucks are involved in getting commodities to warehouses and distribution centres before they make that journey. As reported in *Logistics Management*, 'the role of air cargo has been huge in the growth of e-commerce' (Bentz 2021). 'Given that much of the e-commerce spike is now likely to become permanent', speculated a trucking industry analyst, 'this situation could be great for the trucking industry' (in Bentz 2021). This is a wasteful and damaging allocation of resources. Broader societal ambitions to move away from fossil fuel consumption are undermined by such developments.

Amazon (2020) acknowledges that air cargo is 'a highly carbon-intensive part of the global supply chain'. Yet, rather than seeking alternatives to air cargo, it purchased 'sustainable aviation fuel' that purportedly 'has the power to reduce carbon emissions by up to 20 percent' (Amazon 2020). Given Amazon's aggressive pursuit of growth across its businesses, we can anticipate the applicability of the Jevons paradox. If these purported innovations and efficiencies in aviation fuel were actually achieved and utilised across the company's fleet (neither of which should be assumed), the ongoing expansion to the fleet, air cargo routes, and various forms of production and consumption ostensibly would cancel out gains.

Under capitalism, there cannot be a point at which Amazon or any corporation is satisfied that it has achieved enough growth. The spectacular growth of Amazon has indeed been accompanied by a spike in emissions. As reported by *CNBC*, 'Amazon's carbon emissions jumped 18% [in 2021]' even as 'Amazon lowered its carbon intensity, which measures emissions per dollar of sales' (Palmer 2022).

Consumer desire for convenience is likewise defined by a unidirectional logic of increase. There is no point at which enough 'convenience' is attained, as the mass adoption of generation after generation of 'time-saving' technologies attests. Instead, as Shove (2003: 196) explains, 'convenient solutions generate demand for more convenience'. The environmental implications are worrying, as 'the cumulative effect is to engender and legitimize new, typically more resource-intensive, conventions and expectations' (Shove 2003: 183). The reasons that such expectations become entrenched is inextricably linked to the sociotemporal order. Demand for convenience can be understood as a counterpart to 'a contemporary sense of always being short of time' (Shove 2003: 172). New technologies and the conveniences they afford have undoubtedly shaped and been shaped by social life in complex ways, for instance, transforming aspects of domestic labour (but not the gender relations underpinning them) (see Wajcman 2015). My focus here is on how we can understand acceleration as an intrinsic characteristic of capitalism.

Efficiency gains at work and in the home tend to be cancelled out by expectations to be more productive and to do more. Hartmut Rosa (2013: 151) sees 'social acceleration' as a defining feature of modernity, and identifies three distinct forms of acceleration that exist in a 'self-reinforcing "feedback system"'. *Technological acceleration* is the '*intentional acceleration of goal-directed processes*', including in transportation, communication, and production (Rosa 2013: 74; emphasis in original). The *acceleration of social change* encompasses changes in 'attitudes and values as well as fashions and lifestyles, social relations and obligations . . . as well as forms of practice and habits' (Rosa 2009: 83). Finally, the *acceleration of the pace of life* speaks to the experience of not having enough hours in the day to get through our to-do lists, and is premised on the claim that people must undertake a greater 'number of episodes of action . . . per unit of time' (Rosa 2013: 79) – be it at work or at home. Technological acceleration and the acceleration of the pace of life in the context of a capitalist system help explain the insatiable appetite for convenience and time-saving commodities. What e-commerce offers – the ability to purchase items anytime, anywhere via mobile devices (McGuigan and Manzerolle 2015) – means consumers can shop *when they have time*. What is more, with two-day, one-day, and even same-day delivery, consumers are freed from the need to plan ahead. Last-minute purchases can be made with the expectation of delivery now.

Conclusions

This chapter has explored how e-commerce and supporting logistics systems and infrastructures help lubricate exchange, distribution, and consumption, setting off more production in an endless, accelerating loop. Big Retail and Big Logistics have acted as agents of acceleration, supporting market exchange, distribution, and, hence, the realisation of value. Amazon has played an especially consequential role in expediting already fast online shopping, using its colossal reach across both retail and logistics to reconfigure the rhythm of contemporary shopping. E-commerce as a sector and Amazon as a powerful corporation serve as potent examples of larger capitalist logics that work to normalise ultimately unsustainable rates of production and consumption. Capitalism abides by a logic of perpetual increase and growth:

> Each time capital passes through the process of production it generates a surplus, an increment in value. It is for this reason that capitalist production implies perpetual growth. This is what produces the spiral form to the motion of capital.
>
> (Harvey 2018: 11)

The bounty of commodities afforded by this extraordinarily productive system has been accompanied by spiralling consumer expectations likewise characterised by a logic of increase (more convenience, more speed). I seek to draw attention to how 'the movement of commodities through space is inseparable from the movement of value through its circuit' (Danyluk 2018: 639) to underline the profound tension between this socio-economic system and planetary boundaries.

Escalating production and consumption and accelerating circulation further strain an already damaged planet, as more material resources and fossil fuels are marshalled to produce and move more things in service of growth. The 'Great Acceleration' (Steffen et al. 2015) initiated in the capitalist world has set off dangerous forms of acceleration in biophysical environments. According to Foster, Clark, and York (2010: 35):

> perhaps the greatest danger of climate change to life is the accelerating tempo of the change in the earth system, overwhelming natural-evolutionary processes and even social adaptation, and thus threatening the mass extinction of species and even human civilization itself.

Harvey (2015: 255) highlights spatial and temporal dimensions of contemporary capitalism's environmental impacts:

> Whereas the problems in the past were typically localised – a polluted river here or a catastrophic smog there – they have now become more regional (acid deposition, low-level ozone concentrations and stratospheric ozone holes) or global (climate change, global urbanisation, habitat destruction, species extinction and loss of biodiversity, degradation of oceanic, forest and land-based ecosystems and the uncontrolled introduction of artificial chemical compounds – fertilisers

and pesticides – with unknown side effects and an unknown range impacts on land and life across the whole planet).

E-commerce and the normalisation of hyper-individualised shopping and delivery enshrine a resource-intensive consumer way of life, whatever the environmental cost.

Notes

1 Walmart developed a partnership with e-commerce company Shopify in 2020 'to expand its third-party marketplace site' (Boyle 2020).
2 See Newsome (2015) for an account of the role of labour in retail logistics.
3 Of course, shopping online does not preclude someone from going to bricks-and-mortar retail stores in addition, or using time 'freed up' to drive elsewhere while undertaking other activities.

References

Adobe (2022) 'Adobe: U.S. consumers spent $1.7 trillion online during the pandemic, rapidly expanding the digital economy', *Adobe: News*, 15 Mar. Available at: https://news.adobe.com/news/news-details/2022/Adobe-U.S.-Consumers-Spent-1.7-Trillion-Online-During-the-Pandemic-Rapidly-Expanding-the-Digital-Economy/default.aspx#:~:text=Growth%20was%20modest%20in%202021,gap%20widened%20to%20%2438.8%20billion

Amazon (2020) 'Promoting a more sustainable future through Amazon Air', *Amazon*, 8 July. Available at: www.aboutamazon.com/news/operations/promoting-a-more-sustainable-future-through-amazon-air

Amazon (2021a) *2020 Amazon annual report*. Available at: https://ir.aboutamazon.com/annual-reports-proxies-and-shareholder-letters/default.aspx

Amazon (2021b) *Amazon: Frequently asked questions*. Available at: https://logistics.amazon.co.uk/marketing/faq

Amazon (2021c) *The beginner's guide to selling on Amazon*. Available at: https://m.media-amazon.com/images/G/01/sell/guides/Beginners-Guide-to-Selling-on-Amazon.pdf?initialSessionID=135-2659779-4989755&ld=SDUSSOADirect&ldStackingCodes=SDUSSOADirect

Amazon (2021d) *Fulfillment by Amazon*. Available at: www.amazon.com/fulfillment-by-amazon/b?ie=UTF8&node=13245485011

Amazon (2021e) 'Press release: Amazon.com announces first quarter results', *Amazon*, 29 Apr. Available at: https://press.aboutamazon.com/news-releases/news-release-details/amazoncom-announces-first-quarter-results-0

Armstrong, G., Kotler, P. and Opresnik, M.O. (2016) *Marketing: An introduction*. 13th edn. Harlow, Essex: Pearson Education.

AWS (2021) *AWS Customer Success*. Available at: https://aws.amazon.com/solutions/case-studies/

Banks, S. and Hajibashi, M. (2021) 'Third party logistics (3PL): More critical than ever', *Logistics Management*, 8 Feb. Available at: www.logisticsmgmt.com/article/ third_party_logistics_3pl_more_critical_than_ever/3pl

Bentz, B.A. (2021) 'E-commerce boom: Welcome to the new reality', *Logistics Management*, 9 Mar. Available at: www.logisticsmgmt.com/article/e_commerce_ boom_welcome_to_the_new_reality/ecommerce

Boyle, M. (2020) 'Shopify advances after deal with Walmart expands its reach', *Bloomberg*, 15 June. Available at: www.bloombergquint.com/business/walmart-partners-with-shopify-to-expand-web-marketplace-business

Brevini, B. (2021a) 'Economic profile', in Brevini, B. and Swiatek, L. *Amazon: Understanding a global communication giant*. London: Routledge, pp. 7–21.

Brevini, B. (2021b) 'Introduction', in Brevini, B. and Swiatek, L. *Amazon: Understanding a global communication giant*. London: Routledge, pp. 1–6.

Buch-Hansen, H. (2019) 'Reorienting comparative political economy: From economic growth to sustainable alternatives', in Chertkovskaya, E., Paulsson, A. and Barca, S. (eds.) *Towards a political economy of degrowth*. London: Rowman & Littlefield, pp. 39–53.

Christopher, M. (2010) 'New directions in logistics', in Waters, D. (ed.) *Global logistics: New directions in supply chain management*. 6th edn. London: Kogan Page, pp. 1–13.

Collins, C. (2021) 'Is e-commerce really sustainable? Understanding its impact on the environment', *Sana Blog*, 17 Feb. Available at: www.sana-commerce.com/ blog/impact-of-ecommerce-on-the-environment/

Danyluk, M. (2018) 'Capital's logistical fix: Accumulation, globalization, and the survival of capitalism', *Environment and Planning D: Society and Space*, 36(4), pp. 630–647. doi:10.1177/0263775817703663

Davis, D. (2020) 'Amazon is the fourth-largest US delivery service and growing fast', *Digital Commerce 360*, 26 May. Available at: www.digitalcommerce360. com/2020/05/26/amazon-is-the-fourth%E2%80%91largest-us-delivery-service-and-growing-fast/

Day, M. and Bloomberg (2021) 'Amazon sales beat estimates as pandemic shopping continues', *Fortune*, 29 Apr. Available at: https://fortune.com/2021/04/29/ amazon-q1-2021-sales-beat-estimates-pandemic-shopping/

Dempsey, H. (2021) 'DHL raises earnings forecast again as shoppers shift online', *Financial Times*, 5 May. Available at: https://www.ft.com/content/452603fa-c1f6-418d-bc0e-e033a6c7db13

DHL (2021) *DHL e-commerce campaign: Keep up with the clicks – 60s*. Available at: https://youtu.be/TnSN42BgT94

Dumaine, B. (2020) 'Amazon was built for the pandemic', *Fortune*, 181(6), pp. 86–92.

Edmondson, J. (2021) 'Investing in ecommerce: the high street has moved online', *IG.com*, 26 Apr. Available at: www.ig.com/uk/investments/news/ investing/2021/04/26/investing-in-ecommerce-the-high-street-has-moved-online eMarketer (2020) 'US ecommerce growth jumps to more than 30%, accelerating online shopping shift by nearly 2 years', *eMarketer*, 12 Oct. Available at: www.emarketer.com/content/us-ecommerce-growth-jumps-more-than-30-accelerating-online-shopping-shift-by-nearly-2-years

eMarketer (2021a) 'Amazon's share of the US ad market surpassed 10% in 2020', *eMarketer*, 6 Apr. Available at: www.emarketer.com/content/amazon-s-share-of-us-digital-ad-market-surpassed-10-2020

eMarketer (2021b) 'Worldwide ecommerce will approach $5 trillion this year', *eMarketer*, 14 Jan. Available at: www.emarketer.com/content/worldwide-ecommerce-will-approach-5-trillion-this-year

Evans, M. (2021) 'Global e-commerce market to expand by $1 trillion by 2025', *Forbes*, 25 Mar. Available at: www.forbes.com/sites/michelleevans1/2021/03/25/global-e-commerce-market-to-expand-by-us1-trillion-by-2025/

Fernie, J., Sparks, L. and McKinnon, A.C. (2010) 'Retail logistics in the UK: past, present and future', *International Journal of Retail & Distribution Management*, 38(11/12), pp. 894–914. doi:10.1108/09590551011085975

Fortune (2021) 'G500: World's largest companies: The list', *Fortune*, Aug/Sept.

Foster, J.B., Clark, B. and York, R. (2010) *The ecological rift: Capitalism's war on the Earth*. New York: Monthly Review Press.

Harvey, D. (2015) *Seventeen contradictions and the end of capitalism*. London: Profile Books.

Harvey, D. (2018) *Marx, capital, and the madness of economic reason*. New York: Oxford University Press.

Hesse, M. (2018) 'The logics and politics of circulation: Exploring the urban and non-urban spaces of *Amazon.com*', in Ward, K., Jonas, A.E.G., Miller, B. and Wilson, D. (eds.) *The Routledge handbook on spaces of urban politics*. Abingdon, Oxon: Routledge, pp. 404–415.

Johnson, B. (2020) 'Prime time: Amazon now earth's biggest advertiser', *AdAge*, 7 Dec. Available at: https://adage.com/article/datacenter/prime-time-amazon-now-earths-biggest-advertiser/2298666

Klein, N. (2019) *On fire: The burning case for a green new deal*. London: Allen Lane.

Manzerolle, V.R. and Kjøsen, A.M. (2012) 'The communication of capital: Digital media and the logic of acceleration', *tripleC*, 10(2), pp. 214–229. doi:10.31269/triplec.v10i2.412

Marx, K. (1992 [1885]) *Capital: A critique of political economy, volume II*. Translated by David Fernbach. London: Penguin Books.

Masters, K. (2020) '6 predictions for Amazon in 2021, and how they will play out for retail brands', *Forbes*, 16 Dec. Available at: www.forbes.com/sites/kirimasters/2020/12/16/6-predictions-for-amazon-in-2021-and-how-they-will-play-out-for-retail-brands/

McAfee, E. (2021) '3 Reasons why Amazon will likely continue to gain e-commerce market share', *Forbes*, 31 Mar. Available at: www.forbes.com/sites/forbesbusinesscouncil/2021/03/31/3-reasons-why-amazon-will-likely-continue-to-gain-e-commerce-market-share/

McGuigan, L. and Manzerolle, V. (2015) '"All the world's a shopping cart": Theorizing the political economy of ubiquitous media and markets', *New Media & Society*, 17(11), pp. 1830–1848.

Moore, J.M. (2015) *Capitalism in the web of life: Ecology and the accumulation of capital*. London: Verso.

Mosco, V. (2014) *To the cloud: Big data in a turbulent world*. London: Routledge.

MWPVL (2021) 'Amazon global supply chain and fulfillment center network', *MWPVL.com*. Available at: www.mwpvl.com/html/amazon_com.html

Newsome, K. (2015) 'Value in motion: Labour and logistics in the contemporary political economy', in Newsome, K., Taylor, P., Bair, J. and Rainnie, A. (eds.) *Putting labour in its place: Labour process analysis and global value chains*. London: Palgrave, pp. 29–44.

OECD (2019) *Unpacking e-commerce: Business models, trends and policies*. Paris: OECD Publishing. doi:10.1787/23561431-en

Palmer, A. (2020) 'Amazon lifts FedEx ground delivery ban for sellers, FedEx shares rise', *CNBC*, 14 Jan. Available at: www.cnbc.com/2020/01/14/amazon-lifts-fedex-ground-delivery-ban-for-sellers.html

Palmer, A. (2022) 'Amazon emissions increased 18% last year as Covid drove online shopping surge, *CNBC*, 1 Aug. Available at: www.cnbc.com/2022/08/01/amazon-says-carbon-emissions-increased-18percent-in-2021.html

Pålsson, H., Pettersson, F. and Hiselius L.W. (2017) 'Energy consumption in e-commerce versus conventional trade channels – Insights into packaging, the last mile, unsold products and product returns', *Journal of Cleaner Production*, 164, pp. 765–778. doi:10.1016/j.jclepro.2017.06.242

Pitts, F.H. (2015) 'Creative industries, value theory, and Michael Heinrich's new reading of Marx', *tripleC*, 13(1), pp. 192–222. doi:10.31269/triplec.v13i1.639

Rosa, H. (2009) 'Social acceleration: Ethical and political consequences of a desynchronized high-speed society', in Rosa, H. and Scheuerman, W.E. (eds.) *High-speed society: Social acceleration, power, and modernity*. University Park, Pennsylvania: Pennsylvania State University Press, pp. 77–111.

Rosa, H. (2013) *Social acceleration: A new theory of modernity*. Translated by J. Trejo-Mathys. New York: Columbia University Press.

Shove, E. (2003) *Comfort, cleanliness and convenience: The social organization of normality*. Oxford: Berg.

Soper, S. (2020) 'Amazon plans to put 1,000 warehouses in suburban neighborhoods', *Bloomberg.com*, 16 Sept. Available at: https://www.bloomberg.com/news/articles/2020-09-16/amazon-plans-to-put-1-000-warehouses-in-neighborhoods

Steffen, W., Broadgate, W., Deutsch, L., Gaffney, O. and Ludwig, C. (2015) 'The trajectory of the Anthropocene: The great acceleration', *The Anthropocene Review*, 2(1), pp. 81–98. doi:10.1177/2053019614564785

Verdon, J. (2021) 'Global e-commerce sales to hit $4.2 trillion as online surge continues, Adobe reports', *Forbes*, 27 Apr. Available at: www.forbes.com/sites/joanverdon/2021/04/27/global-ecommerce-sales-to-hit-42-trillion-as-online-surge-continues-adobe-reports/

Wahba, P (2021) 'Walmart hits speed bump on its way to $75 billion e-commerce goal'. *Fortune*, 17 Aug. Available at: https://fortune.com/2021/08/17/walmart-75-billion-e-commerce-goal/

Wajcman, J. (2015) *Pressed for time*. Chicago: University of Chicago Press.

West, E. (2022) *Buy now: How Amazon branded convenience and normalized monopoly*. Cambridge, MA: MIT Press.

3 Promotional Promises

Fossil Fuel Corporations' Sustainability 'Ambitions'

Introduction

The Intergovernmental Panel on Climate Change (IPCC) has authoritatively concluded that human activity has caused climate change. 'Each of the last four decades has been successively warmer than any decade that preceded it since 1850' (IPCC 2021: 5), and without massive cuts to greenhouse gas (GHG) emissions, temperatures will continue to rise, exceeding safe levels. By making ambitious climate pledges – disclosed and promoted in sustainability reports – fossil fuel corporations are presenting themselves as part of the solution. Indeed, it might seem like there is nothing to worry about – that big business and the marvels of markets are up to the challenge of fixing our planet. This chapter problematises the rosy view of corporate 'sustainability', and advances a critique of oil and gas sector 'net zero' 'ambitions' as a type of greenwashing, understood here as 'an umbrella term for a variety of misleading communications and practices that intentionally or not, induce false positive perceptions of an organisation's environmental performance' (Nemes et al. 2022). According to the International Energy Agency (IEA 2021: 12), '[i]n 2020, clean energy investments by the oil and gas industry accounted for only around 1% of total capital expenditure', and while sector-wide investments in such technologies are increasing (IEA 2022: 85), the figures reported are hardly indicative of a turn away from fossil fuels. An investigation published in *The Guardian* revealed that:

> [t]he dozen biggest oil companies are on track to spend $103m a day for the rest of the decade exploiting new fields of oil and gas that cannot be burned if global heating is to be limited to well under 2C.
>
> (Carrington and Taylor 2022)

And yet, sector-wide net zero promises proliferate.

DOI: 10.4324/9781003001621-4

Advanced capitalist societies must rein in those human activities, institutions, and organisations propelling warming in order to wage a response that matches the scale of the problem – at the speed required to avoid the most extreme projections and worst effects of climate disaster. The activities of large business corporations, especially fossil fuel corporations, are of particular consequence:

> As the engines of the modern global economy, large business corporations underpin the production and consumption of an ever-distending cornucopia of products and services. Fossil fuels and natural resource depletion have been crucial components of economic expansion, with energy supply, industrial production, transportation, construction, and forest/agriculture among the sectors with the most significant contributions to overall GHG emissions.
>
> (Wright and Nyberg 2015: 14–15)

Fossil fuel production fuels the manufacture and transportation of all manner of commodities. As noted in the introductory chapter, one study attributed 'nearly two-thirds of historic carbon dioxide and methane emissions' to just 90 'carbon majors' (Heede 2014: 238); investor-owned corporations Chevron (USA), ExxonMobil (USA), BP (UK), and Royal Dutch/Shell ranked first, second, fourth, and sixth, respectively (Heede 2014: 237).

This chapter begins by laying out a theoretical framework that explores the capitalist view of what nature *is for*, bringing together Nancy Fraser's (2021) tripartite conceptualisation of 'Nature' and Max Horkheimer and Theodor Adorno's (2002 [1944]) critique of instrumental reason. As we shall see, contrary to sanguine management perspectives, fossil fuel corporations cannot resolve the fundamental contradiction between environmental sustainability and economic growth. Accelerating growth – the aim of big business – and rapid deceleration of GHG emissions follow conflicting trajectories. The pursuit of economic growth is infinite and the material basis for growth is finite: 'the earth's stock of fossil fuels, in particular, is confined, and the existing stock can only be burnt once. It is irreversible' (Koch 2019: 73). Rather than leaving fossil fuels in the ground to achieve net zero by 2050, the oil and gas sector is advocating the power of efficiency gains and unproven technologies. Furthermore, in initiatives focused on tree planting, 'nature' is expected to serve double duty, driving yet more production and profits while simultaneously 'fixing' the planet. The gap between discourse (including pledges), actions, and investments in the fossil fuel sector in recent years constitutes evidence of greenwashing (Li, Trencher, and Asuka 2022).[1] I will emphasise a temporal dimension of greenwashing – vowing to adopt sweeping changing in the future – which distracts from ongoing fossil fuel production and defuses serious threats to fossil fuel interests in the short term, delaying a genuine energy transformation.

Conceptualising the 'Nature'-Capitalism Relation

How we might understand the relation between 'nature' and 'society' is subject to considerable academic debate (see Foster 2016), the depth and complexity of which lie outside the scope of this chapter. Briefly, crude dualistic/binary understandings of this relation fail to recognise how nature and society are 'co-produced', with 'species mak[ing] environments, and environments mak[ing] species' (Moore 2015: 7). Human and non-human life and natural environments – our histories and our prospects – are fundamentally intertwined. This fact, however, is sidelined by a capitalist view that perpetuates a 'rift between humanity and nature' and obscures the reality of '[t]he world [as] really one indivisible whole' (Foster, Clark, and York 2010: 7). As John Bellamy Foster (1999: 381) explains, Marx used the concept of 'metabolism' to signal 'the complex, interdependent process linking human society to nature'. The notion of a 'metabolic rift' speaks to 'the material estrangement of human beings in capitalist society from the natural conditions of their existence' (Foster 1999: 383). While academic debate continues, the dominant views of states and capitalists accommodate overarching dynamics of capital accumulation and expansion.

Fraser's (2021) eco-critical theory discerns three conceptions of 'Nature' that can be mobilised together in order to unpack divergent understandings of nature and the implications for ecological crises. We might speak of the 'scientific-realist conception' of the biophysical environment as it is studied by climate scientists – what Fraser (2021: 107) terms Nature I. We can also identify the capitalist conception of nature, Nature II, as something that exists outside humans – nature as:

> the ontological other of 'Humanity': a collection of stuff, devoid of value, but self-replenishing and appropriable as a means to the systemic end of value expansion.
>
> (Fraser 2021: 107)

From this standpoint, nature exists for humans – or, more precisely, for capitalists – as resources to be used, consumed, and put to work in the production of commodities. Thus, capitalists measure, manage, and take from nature, applying criteria defined by business interests. Crucially, Nature II is 'by no means a simple fiction or mere idea'; environmental devastation has been brought about by 'the catastrophic hijacking of Nature I by Nature II in capitalist society' (Fraser 2021: 107). Nature II thinking can help us grasp the limitations of corporate responsibility and sustainability. Lastly, Nature III, 'the object studied by historical materialism', sees nature 'as entangled with human history, shaped by and shaping the latter' (Fraser 2021: 107). By attending to history, we see that 'nature' is not a fixed, unchanging 'thing' out there. Humans interact with the biophysical environment, changing the world in which we inhabit. At the same time, those environments change us.

During the capitalist era, societies have sought to dominate this relationship and to control nature, emboldened by a form of modern scientific and industrial thinking that originated in the European Enlightenment. In *Dialectic of Enlightenment*, Horkheimer and Adorno (2002 [1944]) famously advance a critique of the Enlightenment's self-destruction in the twentieth century; a cold, detached, and instrumental mode of thinking found an unfathomably disturbing manifestation in the Holocaust. For Horkheimer and Adorno (2002 [1944]: 1):

> Enlightenment, understood in the widest sense as the advance of thought, has always aimed at liberating human beings from fear and installing them as masters. Yet the wholly enlightened earth is radiant with triumphant calamity.

'Nature' was a powerful source of such fear and, hence, subjected to human mastery: 'what human beings seek to learn from nature is how to dominate wholly both it and human beings' (Horkheimer and Adorno 2002 [1944]: 2). Abstract concepts like property and progress rendered the social and natural world classifiable, calculable, and, hence, (more) controllable. According to Horkheimer and Adorno (2002 [1944]: 11), 'Nothing is allowed to remain outside, since the mere idea of the "outside" is the real source of fear'. Important here, this 'leveling rule of abstraction, which makes everything in nature repeatable' also 'prepared the way' for industry (Horkheimer and Adorno 2002 [1944]: 9). The instrumentalisation of science and technology accelerated capitalist exploitation of nature.

Orthodox economics assimilates the particular into the general, rendering everything commensurable under the principle of exchange. This economic rationality is an expression of instrumental reason – an 'organ of calculation, of planning . . . [that] is neutral with regard to ends; its element is coordination' (Horkheimer and Adorno 2002 [1944]: 69). Vis-à-vis 'nature', this detached, blinkered version of 'reason' can be seen at work in both mass deforestation and in monoculture tree plantations, where trees are rendered means to capital accumulation. For industry and the capitalist, the common denominator of exchange translates people and planet into substitutable units; 'equivalence itself becomes a fetish' (Horkheimer and Adorno 2002 [1944]: 12). So does growth.

Technological advancement has allowed for increasingly productive harnessing of natural resources and workers. Resulting financial rewards have accorded capitalists considerable power and influence. As observed in the 1944 and 1947 preface to *Dialectic of Enlightenment*:

> The increase in economic productivity which creates the conditions for a more just world also affords the technical apparatus and the social groups controlling it a disproportionate advantage over the rest of the population. . . . These powers are taking society's domination over nature to unimagined heights. While individuals as such are vanishing before the apparatus they serve, they are provided for by that apparatus and better than ever before.
>
> (Horkheimer and Adorno 2002 [1944]: xvii)

This system provides an abundance of consumer goods as compensation to workers, improving the lives of many, but at a huge cost: 'the relation between life and production . . . is totally absurd. Means and ends are inverted' (Adorno 2005 [1951]: 15). The individual and the social world – and the biophysical environments that sustain life – must meet the needs of the capitalist mode of production, not the other way around. However, humans cannot, in fact, completely control nature – a fact underlined by the climate crisis. Thus, fear of nature's power persists, and capitalists continue to try to subdue it through science, technology, and industry.

New technology also must conform to capitalist requirements, otherwise its creators will struggle to secure adequate investment. As we saw in Chapter 2, technology is yet another domain for capital accumulation. New technology also provides a means of generating new efficiencies, which, in turn, propel more productivity. Corporate and state actors are counting on such efficiencies to counter climate change and achieve 'net zero', while also fuelling growth. As a brief aside, it is worth remembering that impressive efficiency gains in the current generation of vehicles, appliances, and devices have failed to deliver sufficient emissions reductions thus far.

This fixation on efficiency obscures larger political, economic, and social changes needed to decarbonise. Elizabeth Shove's (2018: 779) critique of energy efficiency identifies two key problems:

> First, . . . efficiency strategies reproduce specific understandings of 'service' (including ideas about comfort, lighting, mobility, convenience etc.), not all of which are sustainable in the longer run. Second, . . . concepts and measures of efficiency depend on 'purifying' and abstracting energy from the situations in which it is used and transformed. Both tendencies obscure longer-term trends in demand and societal shifts in

what energy is for, and both exemplify a particular moment in the history of energy-society relations.

Government policies and corporate initiatives focused on delivering existing standards of comfort and convenience – albeit more efficiently – naturalise what is an historically specific, unsustainable way of life. The 'like with like' comparisons on which efficiency gains are based crowd out more transformative alternatives; for instance, 'there is never any discussion of how efficient a washing line might be, compared with a mechanical dryer' (Rinkinen, Shove, and Marsden 2020: 5). According to Shove and Walker (2014: 53):

> policies that are designed to deliver similar services but with less energy . . . play an important part in reproducing the status quo and in sustaining and legitimizing contemporary material arrangements and practices.

Other ways of being, making, doing, and consuming fade from view as overemphasis on efficiency 'diverts attention away from the project of developing new, non-modern, configurations of nature and society, and of material culture and practice' (Shove 2018: 787).

It is much easier to get capitalists to buy into an environmentalism focused on producing more efficient versions of existing commodities – a new suite of products available for purchase – than one that undermines capitalist production and individualised consumption. Thus, collective transport – non-profit provision through public mass transit systems – is a tougher sell than electric cars. Whereas leaving fossil fuels in the ground immediately is unthinkable, pledging to pursue more efficient extraction and production has a sound business rationale. Like rationality, efficiency has no intrinsic ethic or tendency. Not all forms of efficiency can be characterised as '"good" forms of efficiency, which have at their heart interpretations of service that are consistent with a radically lower carbon society' (Shove 2018: 786). Thus, optimism regarding the power of efficiency *as such* to 'solve' climate change – divorced from broader changes in how we produce, consume, and live – is misplaced.

In sum, a dominant view of nature as resources for capitalist accumulation (Nature II) reproduces the capitalism-nature relation as instrumental and governed by capitalist domination. This view does damaging ideological work, obfuscating the natural cycles and planetary boundaries inherent to biophysical environments (Nature I), and which limit the planet's ability to keep pace with accelerating production and consumption in perpetuity. Furthermore, as workers and consumers, our lives are shaped by and contribute to this system. Disrupting the status quo comes with the threat of greater economic hardship, creating further barriers to action. Instead of exploring alternative socio-economic systems that could accommodate production compatible with planetary boundaries, advanced capitalist societies are placing a massive bet on technology and the 'power of nature itself' (Nature II) to 'save' the planet and us. Next, I will offer a critical overview of sustainability business trends in order to contextualise such approaches.

Promoting Sustainability, Managing Reputation

Corporate environmental activities fall under management strategies that go by a number of names. Corporate social responsibility (CSR) is characterised as 'discretionary socially or ecologically beneficial activities that companies undertake to benefit society (c.f., Carroll 1979)' (Waddock and Googins 2011: 25). CSR can be distinguished from corporate responsibility (CR) or corporate citizenship (CC):

> CC/CR directly engages the company's business model and can be defined as companies living up to clear constructive visions and core values consistent with those of the broader societies within which the company operates.
>
> (Waddock and Googins 2011: 25)

In the corporate websites of Big Oil corporations, such activities are simply labelled 'sustainability'. Business sustainability refers to 'the ability of firms to respond to their short-term financial needs without compromising their (or others') ability to meet their future needs' (Bansal and DesJardine 2014: 71).[2] Whatever term we use, it is clear that many corporations purport to care about and pursue more than financial success alone, as '[d]escribing profit maximization as the only corporate goal no longer seems legitimate' (Sandoval 2015: 619). The proliferation of CSR, business sustainability, and the like reflects a longer-term shift:

> at least since the 1970s, there have been growing calls around the world for corporations to 'behave' more honourably, and consider more than just profit maximisation or market share as key performance indicators.
>
> (Khamis 2020: 84)

Not responding to growing public pressure to be a good corporate citizen would risk increased scrutiny and reputational damage.

These corporate activities and discourses contribute to the larger promotional apparatus corporations use to maintain economic and social power.

For instance, corporate advertising spending by five major oil and gas corporations between 1986 and 2015 was estimated at nearly $3.6 billion (Brulle, Aronczyk, and Carmichael 2020: 99). A study conducted by Robert Brulle, Melissa Aronczyk, and Jason Carmichael (2020: 97) identified key factors influencing advertising expenditures:

the most powerful and consistent determinants of corporate promotional spending by major oil corporations are a congressional activity on climate change and media coverage of the issue.

Creating an image of responsibility through advertising helps pre-empt tighter environmental regulations; a more aggressive approach is adopted when threats to business-as-usual increase (Brulle, Aronczyk, and Carmichael 2020: 99). Massive spending by 'oil majors' on lobbying against stricter climate policies (Laville 2019) also feeds into this pre-emptive strategy, as do the overarching CSR and business sustainability strategies to which advertising, lobbying, and public relations contribute. The appearance of responsibility projected through CSR serves as a means of maintaining a 'social license to operate' (Miller 2018: 22).

Oil and gas companies invest in promotional communication and persuasion tactics as means of safeguarding the bottom line at the expense of the planet. At the same time, the management concept of the 'Triple Bottom Line' (People, Planet, and Profit) keeps 'gaining popularity and . . . has become part of everyday business language' (Kraaijenbrink 2019), perpetuating the problematic idea that these aims are not contradictory. This Triple Bottom Line framework creates an analogy between financial performance and social/ethical and environmental indicators, asserting that corporations ought to manage and measure social, environmental, and economic impacts. According to business ethics critics Wayne Norman and Chris MacDonald (2004: 243), its advocates believe that non-economic performances 'should be measured, calculated, audited and reported', consistent with '[o]ne of the more enduring clichés of modern management [which] is that "if you can't measure it, you can't manage it"'. From this perspective, capitalists can appreciate the value of 'the environment' only after they have created a corresponding accounting system. Reproducing Fraser's (2021) Nature II, the Triple Bottom Line framework conceptualises the biophysical environment as something to be measured, managed, and used by the capitalist, translated into and ultimately reduced to business interests – even if focused on reputation management and not profits as such.

While corporations may proclaim that people, planet, and profit all matter, they do not matter equally. Interestingly, John Elkington (2018), who introduced the Triple Bottom Line term in 1994, has criticised how it has actually been taken up in practice:

> Whereas CEOs, CFOs, and other corporate leaders move heaven and earth to ensure that they hit their profit targets, the same is very rarely true of their people and planet targets. Clearly, the Triple Bottom Line has failed to bury the single bottom line paradigm.

This is not surprising. Under capitalism, social and environmental performance are necessarily subordinated to economic performance; remaining a viable business is contingent on financial success.

The idea that we can have it all – that capitalism can exist in harmony with the planet – bolsters a promotional discourse designed to produce an image of corporate responsibility and sustainability. Indeed, reporting on sustainability has an important promotional function:

> To sustain and even enhance their reputations, many companies feel the need to report regularly on environmental, social, and governance (ESG) matters, and to do so in ways that are consistent and comparable to what other companies are reporting.
>
> (Waddock and Googins 2011: 36)

Although used in the management of individual corporate brands, sustainability trends can envelop entire sectors and industries, as we see with net zero pledges.

'Sustainability' as a management term obfuscates, collapsing together two very different and, in my understanding contradictory, sets of objectives. *Economic* sustainability addresses issues such as the ability of 'organisations and their business models . . . to create and sustain jobs for workers and generate surplus value in their production process', whereas *ecological* sustainability 'refers to the planetary carrying capacity for humanity, and how organizations and society impact on this' (Hill and McDonagh 2021: 46). As Tim Hill and Pierre McDonagh (2021: 46; emphasis in original) explain, 'business and organizations now use sustainability in both of these ways', with corporations positioning themselves as 'ever-willing and useful *solution providers*'. Corporations and their CEOs share their vision for how best to respond to the societal challenge of climate change in order to avoid alternatives that could be bad for business.

For big business, sustainability reporting is standard practice. According to the 2020 iteration of KPMG's sustainability survey, at least 90 percent of Fortune's 250 largest corporations engaged in sustainability reporting from 2011 to 2020, with 76 percent stating carbon reduction targets in

2020 (KPMG IMPACT 2020: 10, 41). The oil and gas sector was among the industry leaders in sustainability reporting rate, and 78 percent of oil and gas companies linked their sustainability goals to the UN Sustainable Development Goals (SDGs) (KPMG IMPACT 2020: 16, 44, 46). Briefly, the SDGs are part of the United Nations Department of Economic and Social Affairs' 2030 Agenda for Sustainable Development, which 'provides a shared blueprint for peace and prosperity for people and the planet, now and into the future' (UN DESA n.d.). This blueprint is committed to 'spur[ring] economic growth – all while tackling climate change and working to preserve our oceans and forests' (UN DESA n.d.). Economic growth, climate action, biodiversity, and responsible consumption and production are presented as mutually compatible (UN DESA n.d.). This view does not stand up to closer scrutiny. Reviewing evidence and modelling for de-coupling economic growth from negative environmental impacts, economic anthropologist Jason Hickel (2019) found that impacts from indiscriminate economic growth outweigh projected resource efficiency improvements. According to Hickel (2019: 881; emphasis in original):

> achieving the sustainability objectives of the SDGs requires that we rethink aggregate global economic growth as a development strategy. The human development objectives of the SDGs can be more safely and feasibly achieved by shifting a portion of global income from richer nations to poorer nations. In other words, reducing *global* income inequality becomes the only reasonable method by which the SDGs can accomplish the human development objectives without violating the sustainability objectives.

Important for my purposes here is how referencing SDGs allows fossil fuel corporations to forge an association that gives the impression of responsibility, sustainability, and legitimacy.

Sustainability priorities are strategic and corresponding reports are selective. KPMG IMPACT's (2020: 49) corporate survey revealed that, unsurprisingly, corporations most prioritised the sustainable development goal focused on economic growth and employment, and while climate change was the second most prioritised SDG, biodiversity-focused SDGs were the least prioritised. In this instance, a corporate sustainability trend equates environmental sustainability with 'climate change' and not 'biodiversity'. The blinkered, instrumental approach adopted by such corporations breaks complex natural systems into divisible items measured and managed on a ledger. Because the metric used does not account for the *relationship between* climate change and biodiversity, the latter does not 'count' towards these climate goals and is rendered less valuable. Additionally, KPMG found that corporations'

SDG reporting 'focuses almost exclusively on the positive contributions companies make towards achieving the goals, and lacks transparency of their negative impacts' (KPMG IMPACT 2020: 48). This selectiveness underlines how such reports serve as tools of reputation management.

The overarching corporate sustainability discourse to which sustainability reports contribute, which collapses together economic and environmental sustainability, produces promotional effects. Following Andrew Wernick (1991: 182), we can understand promotion as:

> a complex of significations which at once represents (moves in place of), advocates (moves on behalf of), and anticipates (moves ahead of) the . . . entities to which it refers.

Sustainability reports represent corporations (as organisations and brands, including their 'core values'), advocate for corporations (by highlighting positive contributions to communities, 'nature', and shareholders), and anticipate and endeavour to deflect criticism (by projecting an image of responsibility). If we see greenwashing as a practice that 'range[s] from slight exaggeration to full fabrication' (Nemes et al. 2022), and if we recognise that encouraging unduly positive appraisals of environmental performance may or may not be the organisation's intention, we can view dominant approaches to business sustainability as a form of greenwashing.

Irrespective of what a corporation intends, the strength of its environmental actions is constrained by capitalism's growth imperative. As Toby Miller points out, 'both CSR and public policy lack a fundamental critique of capitalist growth as an ideology. Sustainability has become a watchword for compromise' (2018: 23). As Graham Murdock and Benedetta Brevini (2019: 76) assert:

> Capitalism's relentless drive for ever-expanding accumulation is presented in official discourse as the indispensable precondition for economic stability and social 'progress'. This key ideological support for business as usual needs to be vigorously contested.

The corporate version of sustainability reinforces the naturalisation of the relentless expansion of production and consumption.

According to Horkheimer and Adorno (2002 [1944]: 118), 'Ideology becomes the emphatic and systematic proclamation of what is. . . . The mere cynical reiteration of the real is enough to demonstrate its divinity'. Sustainability reports reiterate and project into the future the capitalist way of life as 'what is', giving the impression of its inevitability. The net effect of CSR, CR, CC, the 'Triple Bottom Line', and similar management and marketing trends, is 'ideologically propping up globalised capitalism at a time of crisis' (Hill and McDonagh 2021: 49).

Net Zero . . . by 2050?

In contemporary sustainability trends, 'net zero' has surfaced as a government and industry watchword. According to the United Nations (UN n.d.):

> Put simply, net zero means cutting greenhouse gas emissions to as close to zero as possible, with any remaining emissions re-absorbed from the atmosphere, by oceans and forests for instance.

This explanation emphasises deep cuts as the primary strategy, but the term itself is open to alternative interpretations. As Oxfam observes:

> it is striking how much that one small word 'net' can obscure. 'Net zero emissions' and 'zero emissions' do not mean the same thing. . . . [I]nstead of focusing primarily on the hard work of cutting emissions . . . [net zero targets] rely instead on using other methods to remove carbon from the atmosphere. This can allow countries and corporations to continue to pollute, as the millions of tonnes of carbon emissions their factories and powerplants produce will somehow then be removed from the atmosphere, cancelling out their pollution and supposedly achieving 'net' zero.
>
> (Sen and Dabi 2021: 6, 7)

The flexibility and ambiguity of the idea helps explain its popularity among industry actors.

Organisations across the globe, in the spheres of both government and business, are developing plans for decarbonisation by 2050, in line with aims laid out in the Paris Agreement. As reported in an Oxfam briefing paper, 'Of the world's 2,000 largest public companies, . . . at least one-fifth now have net zero commitments' (Sen and Dabi 2021: 9). Given their need to keep growing profits and, hence, production, many business corporations appear more interested in purchasing carbon offsets and exploring the potential of emissions removal technologies than in pursuing massive emissions *reductions*. According to the Global Energy and Climate Innovation Editor at *The Economist*:

> Most companies plan to make some cuts in emissions . . . and to achieve the rest of their carbon goal by buying cheaper 'offsets' (such as credits for renewable energy projects or for protecting forests that sequester

carbon) that vary wildly in quality. This loophole lets firms make green promises without explaining exactly how they plan to clean up their act.

(Vaitheeswaran 2021)

Championing new technology and the use of 'nature' in offsetting strategies helps delay the more difficult challenge of making drastic cuts to emissions while remaining profitable.

The climate 'solutions' offered by Big Oil are expressed through targets framed as 'ambitions', 'aims', and 'aspirations', which raises questions about how accountable these companies must be for meeting them. BP's (2022: 19) sustainability report, titled *Reimagining Energy for People and Our Planet*, states that the company's 'ambition is to be a net zero company by 2050 or sooner, and to help the world get to net zero'. Shell's (2022: 21) report, *Responsible Energy*, communicates its strategy and targets:

> Our climate target is to become a net-zero emissions energy business by 2050, in step with society's progress in achieving the goal of the UN Paris Agreement on climate change.

Advancing Climate Solutions: 2022 Progress Report expresses ExxonMobil's (2022: 6) aim 'to achieve net-zero Scope 1 and 2 greenhouse gas emissions from its operated assets by 2050', the former being 'direct greenhouse gas emissions from Company Operations' and the latter being 'indirect greenhouse gas emissions from energy purchased by the Company' (ExxonMobil 2022: 9). Chevron's (2021) *Climate Change Resilience: Advancing a Lower Carbon Future,* ConocoPhillips' (2022) *Plan for the Net-Zero Energy Transition*, Eni's (2021) *Eni for 2020: Carbon Neutrality by 2050*, and Total's (2020) *Getting to Net Zero* likewise communicate 'ambitions' or 'aspirations' to reach net zero by 2050. The aggressiveness and earnestness with which these individual corporations will pursue such targets may vary, and some of these oil and gas majors were much more resistant than others to pursuing net-zero targets, announcing plans only recently. Overall, the act of reporting such targets at once discloses and promotes corporations' stated values, activities, and plans, and gives an impression of sector-wide action.

It should be noted that inclusion within such reports of disclaimers regarding 'forward-looking statements' (forecasts often accompanied by hedging words such as 'aims', 'intends', 'plans', etc.) clarifies that targets will not necessarily be reached. As these legal provisos specify:

> Forward looking statements involve risk and uncertainty because they relate to events and depend on circumstances that will or may occur in

the future and are outside the control of [the corporation]. Actual results or outcomes may differ from those expressed in such statements.

(BP 2022: 58)

Such statements cover corporations' net-zero and sustainability targets, hence aims promoted are not binding.

Oil and gas corporations' vision of what lower carbon societies will actually look like remains opaque, but one thing is clear: these companies assume growing energy 'needs' or demand in the meantime. Indeed, an advertisement connected to Chevron's (2022) 'lower carbon' strategy expresses striking passivity regarding its vision and ability to effect change:

> What's on the horizon? The answers lie beyond the roads we know. We recognise that energy demand is growing, and the world needs lower carbon solutions to keep up. At Chevron, we're working to find new ways forward through investments and partnerships in innovative solutions. . . . We may not know just what lies ahead, but it's only human to search for it.

The solutions listed in the advertisement are carbon capture as well as using cow waste and hydrogen as energy sources (Chevron 2022). In its carbon neutrality report, Eni (2021: 6) stresses a sector-wide call 'to respond to growing energy needs while limiting greenhouse gas emissions in order to contribute to the global decarbonisation process'. Unsurprisingly, the focus of the energy industry is not on reducing energy dependence. According to Shell (2022: 62), '[t]he most ambitious scenarios show that as the energy system transitions, the world will continue to need oil and gas for decades'. Sector-wide investment in oil and gas production continues.

It is with good reason that critical climate scientists unmask existing net zero policies, broadly, as a 'dangerous trap':

> Current net zero policies will not keep warming to within 1.5° because they were never intended to. They were and still are driven by a need to protect business as usual, not the climate. If we want to keep people safe then large and sustained cuts to carbon emissions need to happen now. . . . The time for wishful thinking is over.

> (Dyke, Watson, and Knorr 2021)

A type of denial arguably unites many governments and corporations, leading to an over-reliance on optimistic bets on new technology. Rather than pursue huge cuts now, advanced capitalist societies are cleaving to hopes for remarkable contributions of carbon dioxide removal technologies, for instance. As solution providers, corporations offer *commodity solutions* to ecological crises. According to a video connected to ExxonMobil's (2021) 'Advancing Climate Solutions' strategy:

> ExxonMobil is introducing a carbon capture and storage concept so big it's like removing 20 million cars off the road big. . . . Because to help address climate change, it's important to think big.

Such thinking is arguably rather small in terms of the range of alternatives considered. If the goal were to remove 20 million cars from the road, public transport systems and better infrastructure for cycling and walking would be the focus. Corporations are ill-equipped to address systemic causes and to offer solutions that are not about opening up new terrain for capital accumulation.

Importantly, this vision of (capitalist) technology as solution very well may fail on capitalism's own terms, as producers may struggle to find investors for unproven technologies. An interest in short-term returns on investment exists in contradiction with the needs to sustain investment in innovation in the longer term. According to a geologist at financial think tank Carbon Tracker:

> For an oil producer, achieving net zero across upstream production . . . requires either the significant deployment of CDR [carbon dioxide removal] technologies or the purchase of offsets, or both. The deployment of carbon capture, usage and storage at the scales assumed in many scenarios remains a challenging prospect to say the least, and *generally continues to lack a clear value proposition for investment.*
>
> (Coffin 2020; emphasis added)[3]

In other words, the technologies that many net-zero targets bank on are not safe financial bets. As long as capitalists are at the forefront of the energy transition, the task of creating technologies capable of removing greenhouse gas emissions from the atmosphere at the required scale and rate – while turning a profit – will remain. This is a tall order. With capitalist technology (not technology per se), investors expect a payout now for speculative technologies that may produce environmental dividends later.

More broadly, under capitalism, 'technical fixes' are bound by an economic horizon, and 'capital seeks to play a shell game with the environmental problems it generates, moving them around rather than address root causes' (Foster, Clark, and York 2010: 74). Even if new technologies were able to efficiently and profitably reduce *and* remove carbon dioxide, we would face yet another quandary: intensified technological dependence. Indeed, it is worth bearing in mind Horkheimer's (2004 [1947]: 66) observation that '[t]he more devices we invent for dominating nature, the more must we serve them if we are to survive'. On all counts, relying on technology as 'saviour' is a massive gamble.

Nature as 'Solution'

Within net-zero 'ambitions', nature itself is pegged by some fossil fuel corporations as another solution: nature *as* carbon sink. 'Nature-based solutions' and 'land-based removal methods' include approaches such as: 'Enhancing carbon sequestration in forests' ('[p]rotecting existing natural forests, restoring degraded forests and improving forest management'); 'Afforestation/reforestation' ('planting forests on lands where they did not previously grow' and 'planting forests in areas that previously had forests', respectively); and 'Bioenergy with carbon capture and storage' ('burning biomass for energy and then capturing and storing the carbon before it is released back into the atmosphere', which can involve 'plantations of fast-growing trees or grasses to be burned in power plants') (Sen and Dabi 2021: 13). This diverse range of approaches all operationalise Nature II, which is, again, used as 'a tap for production's inputs and as a sink for disposing its waste' (Fraser 2021: 100). Even when the corporation assumes the mantle of nature's saviour rather than plunderer, nature remains positioned as humanity's Other (Nature II), which needs to be managed, controlled, classified, measured, and used – perhaps more efficiently – by humans. There is an extraordinary difference between leaving old-growth forests alone, letting them abide by their own natural rhythms, and producing fast-growing tree plantations, designed and controlled by humans. The former sustain complex ecosystems and biodiversity. The latter are a type of monoculture. Both are assimilated under the concept of 'nature' – a nature that *is for* humans.

Tree planting has become fashionable in the oil and gas sector. A promotional video released by Shell (2020) states that '[t]rees are vital in the fight against climate change. Shell is harnessing nature. Supporting reforestation projects. Protecting forests under threat' (Shell 2020). Shell's CEO has championed 'massive reforestation. Think of another Brazil in terms of rainforest: you can get to 1.5C' (in Vaughan 2018). Tree planting is quantifiable and big numbers – 'millions of trees' promised, in the case of Shell (Gosden 2019) – may have a persuasive promotional effect, but new problems are created by continued reliance on simply moving around capitalism's ecological casualties. Where will space for this Brazil-sized rainforest come from? I will return to this issue shortly.

The foreword to the World Resources Institute and the Nature Conservancy report, *The Business of Planting Trees: A Growing Investment Opportunity*, is explicit about the business potential of tree planting:

> Restoring degraded land has the potential to become big business. Established companies and entrepreneurs are finding new ways to make money from sustainably managed forests and farms. Some are responding to governmental incentives. Others are responding directly to the market, restoring land to generate new products and services, or to differentiate their offerings from the competition. Such enterprises are profiting very nicely by breathing new life into unproductive land.
>
> (Faruqi et al. 2018: 1)

Dominated nature itself is rendered a means of saving the planet while also propelling yet more capital accumulation. But the number of trees planted does not equal the number or variety of trees and other species in a forest. This report highlights a company that uses drones to plant trees in remote locations, with tests suggesting 'success rates ranging from 20 percent in temperate regions to 70 percent in tropical climates' – comparable to manual planting (Faruqi et al. 2018: 18). Indeed, a tree planted is not a tree that thrives or even survives. Ecologist Lourens Poorter views 'a lot of the promises that have been made about planting trees in order to restore forests across the world [as] unrealistic', explains *The Guardian*, given that typically '30%–50% of those trees die, and they only pertain to a couple of species that cannot mimic the natural biodiversity of forests' (Quaglia 2021).

What is measured determines what matters. If a thriving multiplicity of plant species and biodiversity were privileged metrics, appraisal of success would be altogether different but, as it is, these are often tree plantations. Political ecologist Benjamin Neimark (2018) points out that '[p]lantations are not forests', and often replace biodiversity with monoculture and reflect corporations' preference for economically advantageous species. All trees are not equivalent, and a forest is not simply an accumulation of trees. The capitalist view of nature suggests otherwise:

> capital seeks . . . to replace an old-growth forest with all of its natural complexity with a simplified industrial tree plantation that is ecologically sterile, dominated by a single species, and "harvested" at accelerated rates. A detailed "division of nature" thus accompanies the detailed division of labor under capitalism, often with disastrous results.
>
> (Foster, Clark, and York 2010: 203)

Arguably, this detailed 'division of nature' in the timber industry can be observed in the newer tree *planting* industry.

Oxfam estimates that the tree planting and land-based removals proposed by just four oil and gas majors would require an area of land 'double the size of the UK' in 2050 (Sen and Dabi 2021: 20). Again, where will this land come from, and crucially, who will it be taken from? According to Oxfam:

> There is a very real risk that the explosion in net zero commitments will fuel a new surge in demand for land, particularly in low- and middle-income countries, which would lead to mass displacement and hunger.
>
> (Sen and Dabi 2021: 7)

Instrumentalised in multiple ways, nature and land are expected to serve many purposes for humanity at once, but we only have one planet and we can only work it so hard.

What if humanity were to stop interfering with as many vital ecosystems as possible for a sustained period of time? One study found that 'overall, tropical forests can get back to roughly 78% of their old-growth status in just 20 years' (Quaglia 2021). As explained by lead author Poorter, 'Compared to planting new trees, [natural regeneration] performs way better in terms of biodiversity, climate change mitigation and recovering nutrients' (in Quaglia 2021). Such an approach performs less well as a money-making venture, however.

Reducing and removing emissions *and* growing markets are two radically different goals imagined as commensurable under the rubric of business sustainability. The reality is rather different. Fossil fuel corporations' long-term strategies for achieving net zero defer the type of deep structural transformation needed to scale back the resource intensity of human activity and achieve a more balanced relation to the planet. Even if, aided by various new technologies, producers were able to produce fossil fuels more efficiently, it is worth recalling how fossil fuels power an unsustainable way of life underpinned by unsustainable production and consumption more generally. As Max Koch (2019: 74) explains, worryingly:

> greater efficiency in the use of a fossil energy source leads to an increase in demand – not to a decrease – and in fact constitutes a necessary precondition for further capital expansion and economic growth.

The growth imperative compromises the degree to which efficiency gains can be directed towards establishing a less energy dependent way of life.

Corporate promotion of net zero gives the impression of activity and a commitment to change. Perhaps the most damaging persuasive effect of this promotional discourse is its ability to delay short-term action across the business world. The image of busyness and activity – and of corporate willingness to provide 'solutions' – reinforces belief in the sustainability of the capitalist paradigm that is ultimately responsible for the climate crisis. Waiting for a market response and new technologies, and associated lag times, is a risky strategy. We have no time to spare.

Closing Reflections

The logics within advanced capitalist societies presume the inevitability of capitalism and the insatiability of consumer wants, needs, and desires, and hence, the need for a massive system of energy provision. As a result, we have missed opportunities to explore and implement truly transformative ways of making, doing, and being, which might encourage radical *reductions* in energy demand, and hence, a very different approach to the energy transition. Instead, under the guise of 'net zero', oil and gas corporations have been able to smuggle further fossil fuel production into that transition, as more renewables and exciting technology *later* help justify continued fossil fuel production *now*. The 'net zero' discourse places too much emphasis on net zero by 2050, and not enough on the minus 45 percent emissions by 2030 (cuts from 2010 levels) needed to limit global warming to 1.5°C (UN n.d.). Fossil fuel companies' strategic use of this long time horizon is not surprising, given that '[t]his target poses an unprecedented challenge to the energy industries that have played a pivotal role in capitalism's relentless expansion' (Murdock and Brevini 2019: 59). Fossil fuel companies and states share in common the pursuit of growth, and existing commitments by governments fall short of agreed targets. In 2022, the United Nations warned:

> Current national climate plans – for 193 Parties to the Paris Agreement taken together – would lead to a sizable increase of almost 11% in global greenhouse gas emissions by 2030, compared to 2010 levels.
> (UN n.d.).

Governments and corporations continue to delay confronting the contradiction between economic growth and planetary boundaries.

The sleight of hand achieved through corporate 'net zero' hinges on how corporations are able to cash in now in terms of managing reputation, maintaining the 'social license to operate', for actions they promise to take later. We can understand this approach as bolstering more explicit types of greenwashing, the 'corporate strategies – especially discourse and pledges to stakeholders – [that] depict proactive actions that exceed a company's actual environmental performance' (Li et al. 2022). Greenwashing related to climate targets is not necessarily about outright deception in communication so much as creating a positive perception of impressive action by making a flurry of promises about future behaviour. Long timeframes have given corporations an alibi for continuing with business-as-usual in the meantime. This manoeuvre helps pre-empt the type of approach advocated by 41 scientists in their critique of net zero: 'real-zero targets' and 'an international treaty for the termination of fossil fuel production' (Skelton et al. 2020). While management notions of 'business sustainability' claim to resolve the contradiction between the environment and economic sustainability, under capitalism, an economic horizon necessarily limits what corporations can do. The dominant view that perpetual economic growth and high consumption are always desirable and are non-negotiable enables fossil fuel corporations to continue exploiting people and planet for profit, foreclosing serious discussion about radically reducing or stopping fossil fuel production.

A world after capitalism is possible. So too is overcoming the 'ecological rift' (Foster, Clark, and York 2010) and achieving a more balanced relation between humans, other species, and the ecosystems we shape and which shape us. Ever-accelerating capitalist production and energy-intensive consumption are not inevitable. They reflect choices. In the current political and economic order, those with the most power to effect systemic change are more committed to sustaining capitalism than the biophysical environment on which life – and capitalism – depend. The question is how long citizens and consumers will continue to agree to this arrangement.

Notes

1 Given recent legal complaints about fossil fuel advertising as misleading and campaigner calls for it to be banned altogether (Carrington 2021), in the present moment, we might expect oil and gas companies to minimise use of high-profile advertising that may draw attention and, hence, criticism.
2 In reality, present and future business goals are difficult to reconcile, and short-term financial targets take priority. According to Bansal and DesJardine (2014: 76), '[e]ven though sustainability has fast become the fashion, short-termism is increasingly the practice'.
3 'Upstream' refers to 'the exploration, development and production of hydrocarbons', whereas 'downstream' refers to 'the processing of the raw materials into products (e.g. crude oil into petrol/gasoline or petrochemical feedstocks) and subsequent marketing of the products to consumers and businesses' (Coffin 2020).

References

Adorno, T. (2005 [1951]) *Minima moralia: Reflections from damaged life*. London: Verso.

Bansal, P. and DesJardine, M.R. (2014) 'Business sustainability: It is about time', *Strategic Organization*, 12(1), pp. 70–78. doi:10.1177/1476127013520265

BP (2022) *Reimagining energy for people and our planet: BP sustainability report 2021.* Available at: www.bp.com/en/global/corporate/sustainability/reporting-centre-and-archive/archive.html

Brulle, R.J., Aronczyk, M. and Carmichael, J. (2020) 'Corporate promotion and climate change: An analysis of key variables affecting advertising spending by major oil corporations, 1986–2015', *Climatic Change*, 159(1), pp. 87–101. doi:10.1007/s10584-019-02582-8

Carrington, D. (2021) '"A great deception": oil giants taken to task over "greenwash" ads', *The Guardian*, 19 Apr. Available at: www.theguardian.com/business/2021/apr/19/a-great-deception-oil-giants-taken-to-task-over-greenwash-ads

Carrington, D. and Taylor, M. (2022) 'Revealed: the "carbon bombs" set to trigger catastrophic climate breakdown', *The Guardian*, 11 May. www.theguardian.com/environment/ng-interactive/2022/may/11/fossil-fuel-carbon-bombs-climate-breakdown-oil-gas?CMP=Share_iOSApp_Other

Chevron (2021) *Climate change resilience: Advancing a lower carbon future.* Available at: www.chevron.com/sustainability

Chevron (2022) *Exploration: Chevron.* Available at: https://youtu.be/HcJew1WINf0

Coffin, M. (2020) 'BP's net zero ambition: Deciphering the code', *Carbon Tracker*, 14 Feb. Available at: https://carbontracker.org/bps-net-zero-ambition/

ConocoPhillips (2022) *Plan for the net-zero energy transition.* Available at: www.conocophillips.com/company-reports-resources/sustainability-reporting/

Dyke, J., Watson, R. and Knorr, W. (2021) 'Climate scientists: concept of net zero is a dangerous trap', *The Conversation*, 22 Apr. Available at: https://theconversation.com/climate-scientists-concept-of-net-zero-is-a-dangerous-trap-157368

Elkington, J. (2018) '25 years ago I coined the phrase "Triple Bottom Line." Here's why it's time to rethink it', *Harvard Business Review*, 25 June. Available at: https://hbr.org/2018/06/25-years-ago-i-coined-the-phrase-triple-bottom-line-heres-why-im-giving-up-on-it

Eni (2021) *Eni for 2020: Carbon neutrality by 2050.* Available at: www.eni.com/en-IT/just-transition/eni-for-2020.html

ExxonMobil (2021) *Driving climate innovation with carbon capture technology.* Available at: https://youtu.be/GtSoM_vQHfk

ExxonMobil (2022) *Advancing climate solutions: 2022 progress report.* Available at: https://corporate.exxonmobil.com/Climate-solutions/Advancing-climate-solutions-progress-report

Faruqi, S., Wu, A., Brolis, E., Ortega, A.A. and Batista, A. (2018) *The business of planting trees: A growing investment opportunity.* World Resources Institute/The Nature Conservancy. Available at: www.wri.org/research/business-planting-trees-growing-investment-opportunity

Foster, J.B. (1999) 'Marx's theory of metabolic rift: Classical foundations for environmental sociology', *American Journal of Sociology*, 105(2), pp. 366–405. doi:10.1086/210315

Foster, J.B. (2016) 'Marxism in the Anthropocene: Dialectical rifts on the left', *International Critical Thought*, 6(3), pp. 393–421. doi:10.1080/21598282.2016. 1197787

Foster, J.B., Clark, B. and York, R. (2010) *The ecological rift: Capitalism's war on the Earth*. New York: Monthly Review Press.

Fraser, N. (2021) 'Climates of capital: For a trans-environmental eco-socialism', *New Left Review*, 127(Jan/Feb), pp. 94–127.

Gosden, E. (2019) 'Shell branches out into tree-planting campaign', *The Times*, 9 Apr. Available at: www.thetimes.co.uk/article/shell-branches-out-into-tree-planting-campaign-h8r6fljm6

Heede, R. (2014) 'Tracing anthropogenic carbon dioxide and methane emissions to fossil fuel and cement producers, 1854–2010', *Climatic Change*, 122(1), pp. 229–241. doi:10.1007/s10584-013-0986-y

Hickel, J. (2019) 'The contradiction of the sustainable development goals: Growth versus ecology on a finite planet', *Sustainable Development*, 27(5), pp. 873–884. doi: doi:10.1002/sd.1947

Hill, T. and McDonagh, P. (2021) *The dark side of marketing communications: Critical marketing perspectives*. London: Routledge.

Horkheimer, M. (2004 [1947]) *Eclipse of reason*. London: Continuum.

Horkheimer, M. and Adorno, T. (2002 [1944]) *Dialectic of enlightenment: Philosophical fragments*. Translated by E. Jephcott. Stanford: Stanford University Press.

IEA (2021) *World energy investment 2021 – International Energy Agency*. Available at: www.iea.org/reports/world-energy-investment-2021

IEA (2022) *World energy investment 2022 – International energy agency*. Available at: www.iea.org/reports/world-energy-investment-2022

IPCC (2021) 'Summary for policymakers', in Masson-Delmotte, V., Zhai, P., Pirani, A., Connors, S.L., Péan, C., Berger, S., Caud, N., Chen, Y., Goldfarb, L., Gomis, M.I., Huang, M., Leitzell, K., Lonnoy, E., Matthews, J.B.R., Maycock, T.K., Waterfield, T., Yelekçi, O., Yu, R. and Zhou, B. (eds.) *Climate change 2021: The physical science basis. Contribution of Working Group I to the Sixth Assessment Report of the Intergovernmental Panel on Climate Change*. Cambridge and New York: Cambridge University Press, pp. 3–32. doi:10.1017/9781009157896.001

Khamis, S. (2020) *Branding diversity: New advertising and cultural strategies*. London: Routledge.

Koch, M. (2019) 'Growth and degrowth in Marx's critique of political economy', in Chertkovskaya, E., Paulsson, A. and Barca, S. (eds.) *Towards a political economy of degrowth*. London: Rowman & Littlefield, pp. 69–82.

KPMG IMPACT (2020) *The time has come: The KPMG survey of sustainability reporting 2020*. KPMG International. Available at: https://home.kpmg/xx/en/home/insights/2020/11/the-time-has-come-survey-of-sustainability-reporting.html

Kraaijenbrink, J. (2019) 'What the 3Ps of the Triple Bottom Line really mean', *Forbes*, 10 Dec. Available at: www.forbes.com/sites/jeroenkraaijenbrink/2019/12/10/what-the-3ps-of-the-triple-bottom-line-really-mean/

Laville, S. (2019) 'Top oil firms spending millions on lobbying to block climate change policies, says report', *The Guardian*, 22 Mar. Available at: www.

theguardian.com/business/2019/mar/22/top-oil-firms-spending-millions-lobbying-to-block-climate-change-policies-says-report

Li, M., Trencher, G. and Asuka, J. (2022) 'The clean energy claims of BP, Chevron, ExxonMobil and Shell: A mismatch between discourse, actions and investments', *PLOS One*, 17(2). doi:10.1371/journal.pone.0263596

Miller, T. (2018) *Greenwashing culture*. London: Routledge.

Moore, J.M. (2015) *Capitalism in the web of life: Ecology and the accumulation of capital*. London: Verso.

Murdock, G. and Brevini, B. (2019) 'Communications and the Capitalocene: Disputed ecologies, contested economies, competing futures', *The Political Economy of Communication*, 7(1), pp. 51–82.

Neimark, B. (2018) 'Greenwashing: corporate tree planting generates goodwill but may sometimes harm the planet', *The Conversation*, 25 Sept. Available at: https://theconversation.com/greenwashing-corporate-tree-planting-generates-goodwill-but-may-sometimes-harm-the-planet-103457

Nemes, N., Scanlan, S.J., Smith, P., Smith, T., Aronczyk, M., Hill, S., Lewis, S.L., Montgomery, A.W., Tubiello, F.N. and Stabinsky, D. (2022) 'An integrated framework to assess greenwashing', *Sustainability*, 14(8). doi:10.3390/su14084431

Norman, W. and MacDonald, C. (2004) 'Getting to the bottom of "triple bottom line"', *Business Ethics Quarterly*, 14(2), pp. 243–262. doi:10.5840/beq200414211

Quaglia, S. (2021) 'Tropical forests can partially regenerate in just 20 years without human interference', *The Guardian*, 19 Dec. Available at: www.theguardian.com/environment/2021/dec/09/tropical-forests-can-regenerate-in-just-20-years-without-human-interference#:~:text=Tropical%20forests%20can%20bounce%20back,humans%20for%20about%2020%20years

Rinkinen, J., Shove, E. and Marsden, G. (2020) *Conceptualising demand: A distinctive approach to consumption and practice*. London: Routledge.

Sandoval, M. (2015) 'From CSR to RSC: A contribution to the critique of the political economy of corporate social responsibility', *Review of Radical Political Economics*, 47(4), pp. 608–624. doi:10.1177/0486613415574266

Sen, A. and Dabi, N. (2021) *Tightening the net: Net zero climate targets – implications for land and food equity*. Oxfam International. doi:10.21201/2021.7796

Shell (2020) *Nature-based solutions and Shell: New energies*. Available at: https://youtu.be/p-_peqYDtoA

Shell (2022) *Responsible energy: Shell plc sustainability report 2021*. Available at: www.shell.com/sustainability/transparency-and-sustainability-reporting/sustainability-reports.html

Shove, E. (2018) 'What is wrong with energy efficiency?', *Building Research & Information*, 46(7), pp. 779–789. doi:10.1080/09613218.2017.1361746

Shove, E. and Walker, G. (2014) 'What is energy for? Social practice and energy demand', *Theory, Culture & Society*, 31(5), pp. 41–58. doi:10.1177/0263276414536746

Skelton, A., Greiser, C., Fopp, D., Lagerlund, H., Björk, M., Glantz, P. and Carton, W. (2020) '10 myths about net zero targets and carbon offsetting, busted', *Climate Home News*, 11 Dec. Available at: www.climatechangenews.com/2020/12/11/10-myths-net-zero-targets-carbon-offsetting-busted/

Total (2020) *Getting to net zero*. Available at: https://totalenergies.com/getting-net-zero

UN (n.d.) 'For a livable future climate: Net-zero commitments must be backed by credible action', *United Nations: Climate Action*. Available at: www.un.org/en/climatechange/net-zero-coalition

UN DESA (n.d.) 'Do you know all 17 SDGs?', *United Nations Department of Economic and Social Affairs: Sustainable development*. Available at: https://sdgs.un.org/goals

Vaitheeswaran, V. (2021) 'Companies' promises to hit net-zero will be put to the test: The world ahead 2022', *The Economist*, 8 Nov. Available at: https://www.economist.com/the-world-ahead/2021/11/08/companies-promises-to-hit-net-zero-will-be-put-to-the-test

Vaughan, A. (2018) 'Shell boss says mass reforestation needed to limit temperature sies to 1.5C', *The Guardian*, 9 Oct. Available at: www.theguardian.com/business/2018/oct/09/shell-ben-van-beurden-mass-reforestation-un-climate-change-target

Waddock, S. and Googins, B.K. (2011) 'The paradoxes of communicating corporate social responsibility', in Ihlen, Ø., Bartlett, J.L. and May, S. (eds.) *The handbook of communication and corporate social responsibility*. Chichester, West Sussex: Wiley-Blackwell, pp. 23–43.

Wernick, A. (1991) *Promotional culture: Advertising, ideology, and symbolic expression*. London: SAGE.

Wright, C. and Nyberg, D. (2015) *Climate change, capitalism, and corporations: Processes of creative self-destruction*. Cambridge: Cambridge University Press.

4 Overcoming Overconsumption

Closing Reflections and New Directions

Consumer capitalism requires economic growth and high rates of consumption. Change in the opposite direction – radically reducing commodity production and consumption – produces fundamental problems for the smooth functioning of this system. This is the dilemma we face. Avoiding the worst projections of climate catastrophe will require advanced capitalist societies to reimagine and reconfigure how – and *at what pace* – to produce, promote, distribute, and consume material and symbolic goods. To limit global warming to 'well below 2°C' higher than pre-industrial levels (aiming for 1.5°C), as set out in the Paris Agreement (United Nations 2015), transformative, system-wide change must happen fast. As outlined by Oxfam (2021):

> In 2015, governments agreed to the goal of limiting global heating to 1.5°C above pre-industrial levels, but current pledges to reduce emissions fall far short of what is needed. To stay within this guardrail, every person on Earth would need to emit an average of just 2.3 tonnes of CO_2 per year by 2030 – this is roughly half the average footprint of every person today.

To emit less, people must consume less.

However, as we have seen, 'consumption' is a complex and wide-reaching concept. Cuts to some areas of consumption would not be practical, possible, or desirable, and not every person on the planet could or should cut their consumption. Importantly, affluent populations disproportionately contribute to overconsumption and its environmental impacts. As recent 'carbon inequality' research on growing emissions from 1990 to 2015 reveals:

> [N]early half of the total growth in absolute emissions was due to the richest 10% . . . The remaining half was due almost entirely to the contribution of the middle 40% of the global income distribution . . . The

DOI: 10.4324/9781003001621-5

impact of the poorest half . . . of the world's population was practically negligible.

(Kartha et al. 2020: 7)

Those who consume well beyond what they need could consume far less.

Rather than exploring how non-essential consumption might be cut immediately, advanced capitalist societies are wasting time debating whether a 'greener' capitalism and innovative new technologies can preserve our current consumer way of life well into the future – albeit more efficiently. The dominant discourse and accompanying political inertia delay action, forestalling deep emissions cuts in the short-term, by insisting that green transitions can satisfactorily be delivered in a capitalist framework. Thus, political calls for 'green growth' crowd out environmental rallying for alternatives to growth. As 'social acceleration' theorist Hartmut Rosa (2013: 281) observes:

> the mutual escalation of growth and acceleration . . . appears to be an unavoidable structural compulsion that mercilessly propels the development of society and turns the 'zero-growth option' into an unreachable utopia.

As currently configured, the business world and governments depend on growth, without which 'jobs are lost, companies close down, tax revenues decrease, while state-expenditure (for welfare and infrastructural programs) increases' (Rosa 2017: 440). Abandoning the pursuit of growth leads to political, economic, and social crises.

Consumer Society and Ecological Crisis has drawn attention to how affluent capitalist societies are stuck in this juggernaut of escalation, with heavily packaged convenience goods, high-speed delivery via e-commerce-based shopping, and the fossil fuels that feed into both providing potent examples of environmentally harmful dynamics of growth and acceleration. I have explored how marketing, logistics, and distribution infrastructures, together, promote and deliver on the promise of convenience and speed and, by extension, produce excessive waste and unsustainable resource consumption. Marketing assumes a powerful role in bringing about exchange, fuelling the never-ending cycle of production and consumption.

In this concluding chapter, I will return to facets of 'consumption' discussed in the introductory chapter, connecting them to the examples examined in this book: plastic packaging and fast-moving consumer goods, e-commerce and Big Logistics, and fossil fuel corporate sustainability. I also will consider further the role of advertising and our media culture in naturalising this consumer way of life and the presumed inevitability of capitalism. Advertising does important cultural and temporal work, building short-term attachments to or interest in commodities and brands. By fixating on what we as individuals want, like, and desire now, promotional messages crowd out the representation of what we collectively may want, like, and desire tomorrow. Long-term societal and environmental needs are an afterthought in a system that pivots on the individual consumer. This chapter will reflect on the limitations the capitalist system places on 'progress' and on technology. While my view is that system-level change is needed to mitigate ecological crises, I will gesture towards smaller steps that could nonetheless play a role.

Consumption, Commodities, and Materiality

In the introductory chapter, I unpacked what it means to 'consume' – by no means a straightforward notion. By prising open this term, we can distinguish *purchase* and *acquisition* from the actual *use* and *using up* of commodities and, hence, pinpoint ecological issues tied to different moments in the circuit of commodity production, exchange, distribution, consumption, and disposal. Capitalists focus on bringing about commodity exchange and consumer purchases. What people actually do with items purchased and associated problems of waste are far less consequential to the business world.

As we saw in Chapter 1, with plastic packaging in the fast-moving consumer goods (FMCG) sector, the primary commodity, the actual consumer good, is used up (e.g., a beverage is drunk or snack eaten), but the accompanying commodity, the *plastic package*, remains, becoming a problem at the moment of disposal. The package is designed to communicate persuasive branding and to make transport and retail convenient, all of which promote and enable purchase and acquisition. A particular material, *plastic*, and application, *the package*, bring about safe and fast movement and enticing display. The consumer good is produced as ephemeral whereas the package is made to last. Barriers to overcoming the plastic pollution crisis, I suggest, are tied not only to the material form – plastic as durable and environmentally harmful – but, crucially, to the commodity form in the Marxist sense – that is, not as 'mere object' but the 'form that a thing takes when exchanged' (Karatani 2016: 171).

Under capitalism, commodities *are for* exchange, a basic fact that unites mundane commodities, such as packaging, with more meaning-rich commodities, such as new fashions and technologies. Financially viable business is contingent on perpetual purchase and acquisition of commodities, irrespective of their usefulness to the individual or society, or even their harmfulness to the biophysical world. The 'single use' character of FMCG and the plastic package makes them ideal commodities from a capitalist perspective, as repeat purchase is built into this type of convenience-oriented consumption. From an environmental perspective focused on resource consumption, we can see the relentless production of 'disposable' packaging – rather than containers designed for reuse – as hugely inefficient.

Addressing the plastic pollution crisis goes against the business interests of powerful petrochemical *and* FMCG corporations. As with any industry, both seek to grow their markets. With the plastics industry, packaging amounts to 'approximately 40% of global end markets' (Mah 2022: 3). Furthermore, even ostensibly more sustainable products are tied to ongoing plastic production:

> New plastics markets are also rapidly proliferating in green technologies. According to the International Energy Agency (IEA), plastics will be the biggest driver of oil demand in the energy transition, reaching close to half of global oil demand by 2050.
>
> (Mah 2022: 3)

If some of these new applications prove necessary, it is even more the case that non-essential applications would need to be cut in order to compensate for growth in other areas. To slow down oil *and* plastic demand would require shifts in consumer practices: a move away from single-use based, convenient consumption. In societies where significant time pressure characterises everyday life *and* powerful corporations decide which products are made available, this is easier said than done.

E-commerce and Big Logistics are likewise optimised for making purchase and acquisition convenient and fast. Whereas packaging renders commodities more mobile, the digitalisation of retail renders the act of shopping itself mobile. Once payment information has been shared, online purchase of commodities is possible anytime, anywhere with the click of a button, resembling dynamics of time-shifting characteristic of streaming media. Amazon has assumed a central role in remaking temporal expectations in both arenas – shopping and media use. As we saw in Chapter 2, online retail is only part of the process of ensuring successful commodity exchange. I drew attention to the tasks of coordination and delivery assumed by business logistics, which enable acquisition. Again, whether the consumer actually uses/uses up or simply accumulates and stockpiles commodities is not the concern of these companies. Once the commodity reaches the consumer's door, it is no longer the retail or logistics company's problem, unless the consumer wishes to return the product – an aspect of online shopping that further intensifies the resource-intensiveness of individualised courier delivery. What is essential is that people keep buying.

The e-commerce-driven acceleration of shopping is simply the latest in a longer historical shift towards the extraordinary accumulation of stuff. The example of the United States, an affluent nation characterised by especially high levels of consumption, provides a powerful illustration of this trend:

> In 1960 the average person consumed just a third of what he or she did in late 2008. Since 1990, inflation-adjusted per-person expenditures have risen 300 percent for furniture or household goods, 80 percent for apparel, and 15–20 percent for vehicles, housing, and food. Overall, average real per-person spending increased 42 percent.
>
> (Schor 2011: 26)

According to Juliet B. Schor (2011: 37), 'acquisition has been paired with product abandonment' – a byproduct of the extension of the 'fashion model', wherein purchasing is tied to frequent style, trend, and fashion changes well beyond the case of apparel (Schor 2011: 40). All manner of consumer durables have arguably become FMCG, or 'products that move quickly through the market', given the speed-up in their 'life cycle' (Schor 2011: 31).

Commodities are at once material and symbolic, and their material qualities are far more stable than their symbolic content. If consumption brings about issues not only of 'disposal' but also 'divestment' and 'devaluation' (Evans 2019: 507), as discussed in the introductory chapter, the fashion cycle raises all three. By continually changing products and promoting new versions, producers and marketers of consumer goods, in effect, erode longer term connections to and the cultural relevance of commodities. What Schor (2011: 41) terms the 'materiality paradox' captures how, 'when consumers are most hotly in pursuit of nonmaterial meanings, their use of material resources is greatest'. As people consume as a means of communication and symbolic expression – to express identity, assert social status, or forge a sense of belonging – they tend to purchase, acquire, and dispose of goods more quickly, as the social meanings associated with particular brands and fads come and go.

Advertising, discussed below, helps produce *and* expunge the symbolic content of commodities, keeping processes of commodity replacement and, hence, the circuit of production, distribution, exchange, and consumption in motion. The media industries likewise legitimate, glamourise, and, in effect, promote a consumer way of life that over-privileges novelty and change and under-privileges function and durability. The cultural power of advertising alone does not animate the consumer capitalist system, however.

The power of fossil fuels has been essential to the expansion of unsustainable rates of consumption. Without massive energy consumption, this entire system of production could not continue apace. As we saw in Chapter 3, energy may be 'reimagined' insofar as continued fossil fuel extraction is complemented with aspirations for and promotional talk about reaching 'net zero' among fossil fuel companies. What is not reimagined, however, is a way of life contingent on high rates of energy consumption. Instead, the capitalist imagination promises that through new technology and innovation, everything can stay more or less the same. Even 'nature' becomes a site of capitalist innovation, as it is positioned as both commodity and means of economic growth (Fraser 2021) *and* climate change 'solution' via expansive tree planting schemes, for instance. A temporal dimension of greenwashing is expressed in corporate fossil fuel 'net zero' initiatives designed to assure the public that requirements of planet and profit are reconcilable and that climate change can be 'solved' later, hence, there is no need to disrupt business-as-usual now.

Unpacking how we understand consumption, 'a highly ambiguous, polyvalent category' (Rosa 2019: 254), is an important step towards pinpointing the populations and consumer practices most responsible for overconsumption. Addressing the climate crisis will require 'consuming differently' *and* 'not consuming as much. So, volume and composition' (Schor in Stewart 2021). To establish a more just and balanced relation to the planet, we must reckon with the fact that the purchasing behaviour of affluent consumers has taken a disproportionate toll, as Naomi Klein (2019: 122) points out:

> Climate change demands that we consume less, but being consumers is all we know. Climate change is not a problem that can be solved simply by changing what we buy – a hybrid instead of an SUV, some carbon offsets when we get on a plane. At its core, it is a crisis born of overconsumption by the comparatively wealthy, which means the world's most manic consumers are going to have to consume less so that others can have enough to live.

To ensure a fairer or more just '*distribution* of access' (Evans 2019: 511; emphasis in original), decreases in volume by affluent populations will need to accommodate *increasing* consumption in underserved regions and populations. However, consumption by affluent people drives growth, and the commercial media system is premised on selling advertising targeted at these affluent markets. To this media system I now turn.

Advertising, Consumer Culture, and Promotional Culture

Media and communication studies contributions to consumer society critiques assign central importance to advertising for good reason. Advertising, as both 'a system of ideology and a system of media support' (Turow and McAllister 2009: 2), is a powerful tool for translating commodity production into consumption. In addition to glamourising and legitimising consumer desires, advertising 'distorts the reality of harmful production processes by omission' (Park 2021: 54). The joys of consumption are over-represented, and problems tied to production and circulation, including environmental consequences, are under-represented or erased. Moreover, as Jonathan Hardy (2014: 152) points out, '[d]ependence on advertising finance is the single most important way in which advertising influences media content' and is a key factor that 'influences how media markets are organised', resulting in media that principally cater to the affluent consumers sought by advertisers.

Although a diverse range of stories circulate in popular media, taken together, television, magazines, social media, and so on naturalise a vision of happiness or the 'good life' tied to commodity consumption. Advertising, digital media, and journalism '[a]ll encourage overconsumption or are economically supported by commercial organizations that profit from overconsumption' (Park 2021: 2). Correspondingly, economic growth is widely endorsed as necessary and good, even in much journalism. For instance, journalism in the United States and Britain has provided an overwhelmingly unbalanced account of economic growth as positive (Lewis 2013: 123–126). The forms of marketing and logistics examined in this book would not act as *agents of acceleration* if they were not put to work in a culture that takes for granted commodity desires and expectations of growth – assumptions reinforced and amplified by advertising and media. By focusing on behind-the-scenes and more mundane dimensions of marketing, which have not received as much critical attention as advertising, I hope to broaden debates about consumer society that foreground social meaning, desire, and persuasion. However, I recognise the cultural role assumed by marketing in relation to the latter as fundamental to the consumer capitalist system.

The spread of advertising has been tied to the genesis of a culture that overwhelmingly addresses people as consumers and which is characterised by promotional logics. We can understand *consumer culture* as 'the symbols and messages that surround people about products and services that we buy and use (that is, consume)' and 'how people make meaning from these messages' (Turow and McAllister 2009: 4). Such symbols are not limited to advertising. The proliferation of 'branded content' that further complicates and compromises the boundary between advertising and media content (Hardy 2022), branding deals with musicians that render popular music a marketing vehicle (Meier 2017), and corporate sponsorship deals with arts and cultural institutions (including with Big Oil) (Miller 2018) illustrate some of the ever-expanding ways messages about commodities and brands saturate contemporary culture. Drawing on Andrew Wernick (1991: 182), we can understand *promotional culture* as one in which:

> the range of cultural phenomena which, at least as one of their functions, serve to communicate a promotional message has become . . . virtually co-extensive with our produced symbolic world.

Thus, when pondering which commodity to purchase or to invest meaning in, one encounters brands already assigned meanings explicitly by advertisements and in more complex and subtle ways by media and culture more generally.

While advertisements may promote a particular product or brand, the broader consumer culture or promotional culture, in effect, promotes a consumer a way of life. This way of life caters to individual, short-term needs and wants rather than collective or long-term needs and wants and downplays the significance of environmental needs. The system of promotion helps sustain *culturally* an environmentally unsustainable system.

We can understand this backdrop of consumer culture and promotional culture as shaping the ways the plastic bottle recedes behind the positive associations cultivated by powerful brands owned by Coca-Cola, PepsiCo, Unilever, Nestlé, and Procter & Gamble. We not only consume material commodities such as snacks and soaps. We connect to ideas such as youthfulness and beauty. By virtue of this same dynamic, however, negative associations assigned to problematic materials such as plastic can threaten brand reputation, hence why FMCG companies are scrambling to offer more sustainable alternatives and brand themselves as 'eco'. Maintaining the symbolic content of brands requires ongoing promotional efforts and adaptation to – or at least gesturing towards – cultural shifts and social change, including environmental demands. Advertising and packaging, together, contribute to this promotional work.

The sheer volume of self-interested messages circulating in a promotional culture means it becomes difficult to evaluate environmental claims made by companies, separating greenwashing campaigns that exaggerate organisations' environmental efforts from genuine changes in business practices. The flurry of activities and pledges promoted in fossil fuel corporate sustainability reporting and communications makes it harder to grasp the fundamental contradiction between planet and profits under capitalism. Comprehensively highlighting *what* (i.e., listing a panoply of activities and ambitious targets) overshadows a crucial part of climate action: *when* (i.e., the real timescales by which genuine change will be realised). Big promises about a brighter future serve as strategies of deferral, suggesting that nothing major needs to change now.

In fact, e-commerce and high-speed logistics suggest we can have even more *right now*. Online retailers are always open for business, so buying and selling need not stop. Behemoths such as Amazon reap the rewards of a consumerist ideology promulgated in promotional culture that assumes commodity desires are inexhaustible. According to Emily West (2022: 15), 'Amazon in particular wants to be the "everything" brand for "everyone"' – a service 'defined by the convenience and ease of the consumer'. Endless product options, fast and easy transactions, and high-speed delivery all point in one direction: more resource-intensive consumption, as the 'friction that Amazon removes speeds up the circuit of capital, making it more profitable by reducing inventory costs and encouraging more spending' (West 2022: 46). By tightening the feedback loop between production and consumption, e-commerce and Big Logistics expedite the circulation of capital.

The circulation of meaning and materials brought about by marketing and distribution activities keeps commodities moving and selling. Buying new products and developing new technologies seems to express progress. However, this version of progress is constrained by the requirements of the capitalist system and comes with a steep environmental price.

Fettered to Capitalism: The 'Culture Industry', Progress, and the Limits of Technology

Drawing on the Frankfurt School, we can grasp the complex ways capitalism, the profit motive, and the growth imperative condition media, technology, and even the idea of 'progress' in consumer societies. The power of what Max Horkheimer and Theodor Adorno (2002 [1944]) famously termed the 'culture industry' lies not in directly convincing people to think, buy, or do things they otherwise would not. Instead, more significant ideological work relates to 'the emphatic and systematic proclamation of what is' (Horkheimer and Adorno 2002 [1944]: 118) and the corresponding call to 'conform to that which exists anyway' (Adorno 2001 [1963]: 104) – not because it is desirable, but because there is no alternative. According to Horkheimer and Adorno (2002 [1944]: 136), 'the triumph of advertising in the culture industry' is 'the compulsive imitation by consumers of cultural commodities which, at the same time, they recognize as false'. Setting aside the questionable assumption of 'compulsive' behaviour, this powerful critique asserts that people may consume commodities (cultural and otherwise) and follow trends even if they see through the adverting pitch and do not actually 'buy into' the messages, ideas, and ideals promoted. One may well regard never-ending consumption as problematic and undesirable, but the perpetuation of the idea that no other (enjoyable) way of life is possible leaves little option but to conform, at least to some degree.

The 'culture industry' promotes a capitalist ideology yet still can accommodate films, television, and journalism that speak to environmentalist and even anti-capitalist concerns. Powerful indictments of inaction by world elites can circulate, as was the case when youth environmental activist Greta Thunberg (2019) stated at the United Nations Climate Action Summit that '[w]e are in the beginning of a mass extinction, and all you can talk about is money and fairy tales of eternal economic growth'. Even if readers and viewers agree, further discussion may seem futile because the dominant view perpetuated in commercial media is that there is no other viable option. Media industry accounts privilege '"green growth", "sustainable consumption" and . . . "reforming capitalism"' as ways to 'transition to a lower-carbon economy (and society)' (Morgan 2015: 78).

Capitalist markets and new technologies are widely imagined as forces of progress, and as capable of solving mounting environmental problems, or so it is hoped. With a Frankfurt School lens, we can question the substance of notions of 'progress' tethered to consumption and technological innovation, asking who really benefits and at what cost to the planet. Writing in the 1970s, Erich Fromm (2013 [1976]: 1) identified the negative consequences of the version of progress inaugurated under industrial capitalism:

> [T]he promise of domination of nature, of material abundance, of the greatest happiness for the greatest number, and of unimpeded personal freedom . . . has sustained the hopes and faith of the generations since the beginning of the industrial age. . . . [F]rom the substitution of mechanical and then nuclear energy for animal and human energy to the substitution of the computer for the human mind, we could feel that we were on our way to unlimited production and, hence, unlimited consumption; that technique made us omnipotent; that science made us omniscient.

The reality realised contrasts starkly with the promise, as '[e]conomic progress has remained restricted to the rich nations, and the gap between rich and poor nations has ever widened' and 'technical progress itself has created ecological dangers and the dangers of nuclear war' (Fromm 2013 [1976]: 2). Fromm finds the origins of this failed promise not only in capitalism's fundamental economic contradictions, as underlined in this book, but also in a flawed understanding of human psychology: the assumption that life is about satisfying individual desires and needs, and that selfishness is the basis of a harmonious society (Fromm 2013 [1976]: 3).

By placing the wellbeing of humans at the centre of this account, the critique Fromm advances in *To Have or To Be?* (2013 [1976]) complements more recent thinking on 'degrowth', which argues for 'refocus[ing] on what really matters: not GDP, but the health and well-being of our people and our planet' (Kallis et al. 2020: xv). Thinking from the degrowth movement demonstrates that transcending the growth paradigm is not impossible, even as pursuing the 'continuous contraction of the material throughput in economies of rich countries until a sustainable steady state is reached' would face serious challenges and 'require radical change in underlying cultural values' (Büchs and Koch 2019: 162–163). Ways of conceptualising and measuring 'wellbeing' vary in post-growth discourses (Büchs and Koch 2017: 58–67), discussion of which lies beyond the scope of this concluding chapter. These debates provide essential starting points for broadening our understanding of what progress could be.

Building on the Frankfurt School tradition, social theorist Rosa (2017, 2019) also conceptualises the relationship between the failure to deliver wellbeing in modern societies and the compulsion to grow – the latter of which he terms 'dynamic stabilisation'. For Rosa (2017), such societies define the 'good life' in terms of the pursuit of wealth and resources, leading to an orientation to the world that both damages the planet *and* fails to remedy forms of alienation produced by capitalism. An idea of the good life that could overcome this impasse would focus on realising a 'resonant' way of relating to the world, not the stockpiling of stuff; 'resonance' prioritises responsiveness to experience and transformative social encounters rather than accumulation, instrumentalism, and control (Rosa 2017: 450).

While I cannot provide an in-depth explanation of the nuances of this concept here, I would like to gesture towards how it could contribute to a project of overcoming the consumer society model. A politics focused on creating conditions conducive to 'resonance', including alternatives to the growth paradigm that fuels the 'hamster-wheel of modern social life' (Rosa 2017: 438), could be animated by a vision of the good life defined by stronger social relationships, deeper connections to nature and culture, and even more meaningful relationships to the material objects in our lives. While purchasing and acquiring goods may offer 'a *promise of resonance*' (Rosa 2019: 255; emphasis in original), *these acts* do not bring about resonant relationships with the objects bought. However, we can have meaningful relationships to things. Rosa uses examples such as surfboards and skis, but the same can be said of the garments, furniture, and technologies that are increasingly thrown away. Where Schor's 'materiality paradox' points to the problems with weak attachments leading to the frequent disposal of commodities, Rosa's notion of 'resonance' might offer ways into supporting more enduring relationships to those things we do buy and use. A move away from the consumer way of life could be supported by changing the relationship to consumption and associated activities and objects rather than sacrificing consumption altogether. Fostering deeper attachments to what we buy could contribute to reversing dynamics of overconsumption.

What role might technology assume in such efforts? Horkheimer and Adorno's (2002 [1944]) account of technology as domination – as instrument of war, propaganda, and capitalist ideology – offers a powerful counterpoint to assumptions that new technology is a force of human liberation. Not simply a tool, under capitalism technology becomes a tool of a very particular kind: profit generation. The profit motive constrains the possibilities of what technology can be and do:

> How far progress and regression are intertwined today can be seen in the notion of technical possibilities. Mechanical processes of reproduction have developed independently of what they reproduce, and become autonomous. They are considered progressive, and anything that has no part in them, reactionary and quaint. Such beliefs are promoted all the more thoroughly because super-machines, once they are to the slightest degree unused, threaten to become bad investments.
>
> (Adorno 2005 [1951]: 118)

What is technically possible and what is possible to do profitably are altogether different matters. As we saw in Chapter 3, the development and roll-out of carbon dioxide removal technologies is hampered by the profit motive. The capitalist vision of technology is as consumer good *and* means of production for manufacturing yet more commodities. 'Productive consumption' by capitalists (see the introduction and Chapter 2) binds us to the cycle of overproduction and overconsumption that compromises the biophysical environments on which we depend.

The efficiency gains from new technology on which environmental hopes are commonly pinned have in practice been cancelled out by greater consumption. For instance, as observed by historian Frank Trentmann (2017: 672):

> Those predicting in the 1970s that households in advanced, rich societies would become saturated once everyone had a TV, a fridge, and most had a car were wrong because they ignored a fundamental dynamic of consumer societies: standards, norms, technologies and habits change.
>
> For one, efficiency carried with it the in-built dilemma that it saved people money, which could be spent on bigger things or greater comfort. Efficiency and consumption shadowed each other. . . . Fridges became more efficient, but they also doubled in size. . . . Engines were made more fuel-efficient, allowing car makers and owners to expect more horsepower under the hood. The energy used per square foot in

new American homes may have fallen, but the houses themselves ballooned.

Efficiencies that register in terms of *financial resources* have undermined efficiencies defined in terms of *natural resource use*.

Those industries that produce new technologies have an interest in shaping changes to such norms and habits by designing obsolescence into the life cycle of products. As Justin Lewis (2013: 133) points out:

> [The] dominant role [of the media industries] in our culture gives us a definition of progress based on the principle that the shorter the lifespan of objects, the faster we are moving forward. Progress and planned obsolescence are thereby seen as intertwined. This blots out space . . . for imagining other forms of human development.

To safeguard the ecological viability of the planet for current and future generations, we must imagine and explore ways of relating to the world that are not so reliant on the consumption of commodities.

Expanding the Possible: Steps Towards Transformation

By taking a step back and looking at the larger system of capitalist production and consumption, this book observes structural forces that have produced an environmental debt comprising climate, plastic pollution, biodiversity, and other ecological crises. It invites advertising studies students and scholars to grapple with the larger socio-economic forces that govern what advertising is ultimately for. I understand marketing and logistical systems as important means through which commodity production, exchange, distribution, and consumption are accelerated. The pursuit of short-term profits and prioritisation of short-term consumer needs and wants are tied to this dynamic.

The capitalist view naturalises consumer societies as inevitable and everlasting. In relation to the history of human life, the history of capitalism and commodity exchange is actually extremely short. Alternatives are possible. This is not to suggest that we revert to pre-capitalist modes of production. Norms, habits, and ideas change. Human societies can learn from history and are capable of creating more just modes of production that can exist in a more balanced relation to the planet. Untethering the notion of freedom from private property might create space for ideas and experiences of freedom compatible with a transition away from overconsumption – perspectives that see freedom not as 'the freedom to accumulate, but the fact that I have no need to accumulate' (Adorno and Horkheimer 2011: 23). Rather than recreating a more efficient version of existing consumer societies, we could stop engaging in activities that compromise genuine progress, starting with those most widely agreed to have harmful social and environmental effects.

There are many potential entry points into realising a less resource-intensive future. The suggestions I gesture towards are by no means exhaustive and deserve much fuller treatment than I can provide here. Briefly, *governments* could tighten up regulations over big business and roll out product bans. Single-use plastic bans, although in my view not currently substantial enough, are a step in that direction. Contributing to and building on existing campaigns, *consumers* could take more control over what is produced by organising mass boycotts of environmentally destructive products and calling out bad behaviour – online or offline – in order to tarnish brands' reputations. *Producers* could pursue genuine innovation focused on environmental outcomes. Rather than needlessly shipping water around in our drinks, detergents, and so on, manufacturers could shift their investments towards syrups, concentrates, and the like to which consumers could add water in the home. For instance, a study on cleaning products conducted by *Which?* magazine 'found that concentrated products used 75% less plastic packaging and 97% less water' (Denyer 2021: 15). Rather than privileging single-use frameworks for consumption, producers could optimise their products for repeat use.

The media industries could play a more substantial role in showcasing alternatives to consumer-based needs, wants, and desires, including those of a collective and environmental, rather than individual and commodity, character. Media organisations could broaden the political debate, giving more time to arguments for degrowth (Kallis et al. 2020; Büchs and Koch 2017) and eco-socialism (Fraser 2021). Greater interaction between media and civil society organisations could help draw more attention to and generate more discussion about alternative ways of being, making, and doing, which could interrupt the normalisation of overconsumption. This might involve helping us imagine how our system of consumer provision would look if notions of *sufficiency* (Speck and Hasselkuss 2015; Spengler 2016) were seen as legitimate and acceptable to consumers and producers alike. Sufficient consumption can be understood in terms of minimums – 'each individual has at least enough to meet basic needs' – *and* maximums, which are necessary to rein in 'the adverse environmental . . . impacts of the consumption patterns in wealthy countries' (Spengler 2016: 936).

Media organisations also could be more discerning in terms of the companies to which they sell advertising. For instance, *The Guardian* banned the sale of advertising to oil and gas companies (Waterson 2020).[1] Dependence on advertisers as financiers limits how far such an approach can be pursued, however. Illustrating this point, *The Guardian* responded to reader requests that the newspaper go further and 'turn down advertising for any product with a significant carbon footprint, such as cars or holidays' by stating that 'this was not financially sustainable while the media industry's business model remained in crisis' (Waterson 2020). Reader, viewer, and listener insistence that media organisations do not sell advertising to problematic industries could encourage greater commitment to such bans (in order to avoid reputational damage).

Helping audiences imagine alternative ways of living could help embolden *citizens* to question political establishments and social orders committed to the defence of capitalism at the expense of the planet and the populations that suffer the worst consequences of ecological crisis. Political mobilisation led by youth environmental activists (e.g., Fridays for Future) has powerfully expressed problems inhering in a system that caters to today's affluent consumers, despite the harms incurred for poorer populations and future generations. It is arguably easier to imagine future people than future ways of making, being, and doing. Building on the successes of existing movements, political actors, consumers, citizens, and workers might amplify calls to explore genuine alternatives to growth-based societies.

A serious response to ecological crisis will require transformative political economic and social change. Fundamentally, this type of change calls for the development of 'a society that *does not need* to grow, augment and innovate *just to maintain the status quo* or to secure its structural reproduction' (Rosa, Dörre, and Lessenich 2017: 64; emphasis in original). This does not mean that all production and consumption should contract, but rather that innovation and growth should reflect altogether different priorities. From my perspective, such a project might start by working towards reversing the 'absurd' 'relation between life and production' identified by Adorno (2005 [1951]: 15). We should produce to live, and in a much more balanced relation to the biophysical environment. The sectors and commodities examined in this book – fast-moving consumer goods and plastic packaging, e-commerce and Big Logistics, and fossil fuels – deliver convenience and speed, but they fail to free people from the capitalist conditions and time pressures that make such features desirable in the first place. They offer failed solutions that come with a steep environmental cost.

Note

1 Even some advertising agencies have stopped accepting oil and gas accounts, drawing parallels between fossil fuels and tobacco as industries with which it is no longer acceptable to do business (Hsu 2021).

References

Adorno, T. (2001 [1963]) 'Culture industry reconsidered', in Bernstein, J.M. (ed.) *The culture industry: Selected essays on mass culture.* London: Routledge, pp. 98–106.

Adorno, T. (2005 [1951]) *Minima moralia: Reflections from damaged life.* Translated by E.F.N. Jephcott. New York: Verso.

Adorno, T. and Horkheimer, M. (2011) *Towards a new manifesto.* Translated by R. Livingstone. New York: Verso.

Büchs, M. and Koch, M. (2017) *Postgrowth and wellbeing: Challenges to sustainable welfare.* Cham: Palgrave Macmillan.

Büchs, M. and Koch, M. (2019) 'Challenges for the degrowth transition: The debate about wellbeing', *Futures*, 105, pp. 155–165. doi:10.1016/j.futures.2018.09.002

Denyer, M. (2021) 'Shopping guide: Washing-up liquid', *Ethical Consumer*, 191(Jul/Aug), pp. 14–17.

Evans, D.M. (2019) 'What is consumption, where has it been going, and does it still matter?', *The Sociological Review*, 67(3), pp. 499–517. doi:10.1177/0038026118764028

Fraser, N. (2021) 'Climates of capital: For a trans-environmental eco-socialism', *New Left Review*', 127(Jan/Feb), pp. 94–127.

Fromm, E. (2013 [1977]) *To have or to be?* London: Bloomsbury Academic.

Hardy, J. (2014) *Critical political economy of the media: An introduction.* London: Routledge.

Hardy, J. (2022) *Branded content: The fateful merging of media and marketing.* London: Routledge.

Horkheimer, M. and Adorno, T. (2002 [1944]) *Dialectic of enlightenment: Philosophical fragments.* Translated by E. Jephcott. Stanford: Stanford University Press.

Hsu, T. (2021) 'Ad agencies step away from oil and gas in echo of cigarette exodus', *New York Times*, 25 Mar. Available at: www.nytimes.com/2021/03/25/business/media/climate-ad-agencies-fossil-fuels.html

Kallis, G., Paulson, S., D'Alisa, G. and Demaria, F. (2020) *The case for degrowth.* Cambridge: Polity.

Karatani, K. (2016) 'Capital as spirit', *Crisis & Critique*, 3(3), pp. 167–189.

Kartha, S., Kemp-Benedict, E., Ghosh, E., Nazareth, A. and Gore, T. (2020) *The carbon inequality era: An assessment of the global distribution of consumption emissions among individuals from 1990 to 2015 and beyond.* Joint Research Report. Stockholm Environment Institute and Oxfam International, Stockholm and Oxford. doi:10.21201/2020.6492

Klein, N. (2019) *On fire: The burning case for a green new deal.* London: Allen Lane.

Lewis, J. (2013) *Beyond consumer capitalism: Media and the limits to imagination.* Cambridge: Polity.

Mah, A. (2022) *Plastic unlimited: How corporations are fuelling the ecological crisis and what we can do about it.* Cambridge: Polity.

Meier, L.M. (2017) *Popular music as promotion: Music and branding in the digital age.* Cambridge: Polity.

Miller, T. (2018) *Greenwashing culture.* London: Routledge.

Morgan, T. (2015) 'Growing ourselves to death? Economic and ecological crises, the growth of waste, and the role of the media and cultural industries', *Human Geography*, 8(1), pp. 68–81. doi:10.1177/194277861500800105

Oxfam (2021) 'Carbon emissions of richest 1% set to be 30 times the 1.5°C limit in 2030: Press release', Oxfam.org.uk, 5 Nov. Available at: https://www.oxfam. org.uk/media/press-releases/carbon-emissions-of-richest-1-set-to-be-30-times-the-15c-limit-in-2030/

Park, D.J. (2021) *Media reform and the climate emergency: Rethinking communication in the struggle for a sustainable future*. Ann Arbor, Michigan: University of Michigan Press.

Rosa, H. (2013) *Social acceleration: A new theory of modernity*. Translated by J. Trejo-Mathys. New York: Columbia University Press.

Rosa, H. (2017) 'Dynamic stabilization, the Triple A. approach to the good life, and the resonance conception', *Questions de Communication*, 31, pp. 437–456. doi:10.4000/questionsdecommunication.11228n

Rosa, H. (2019) *Resonance: A sociology of our relationship to the world*. Translated by J.C. Wagner. Cambridge: Polity.

Rosa, H., Dörre, K. and Lessenich, S. (2017) 'Appropriation, activation and acceleration: The escalatory logics of capitalist modernity and the crises of dynamic stabilization', *Theory, Culture & Society*, 34(1), pp. 53–73. doi:10.1177/0263276416657600

Schor, J.B. (2011) *True wealth: How and why millions of Americans are creating a time-rich, ecologically light, small-scale, high-satisfaction economy*. New York: Penguin.

Speck, M. and Hasselkuss, M. (2015) 'Sufficiency in social practice: Searching potentials for sufficient behavior in a consumerist culture', *Sustainability: Science, Practice and Policy*, 11(2), pp. 14–32. doi:10.1080/15487733.2015.11908143

Spengler, L. (2016) 'Two types of "enough": Sufficiency as minimum and maximum', *Environmental Politics*, 25(5), pp. 921–940. doi:10.1080/09644016.2016.1164355

Stewart, E. (2021) 'Why do we buy what we buy? A sociologist on why people buy too many things', *Vox*, 7 July. Available at: www.vox.com/the-goods/22547185/consumerism-competition-history-interview

Thunberg, G. (2019) 'Transcript: Greta Thunberg's speech at the U.N. Climate Action Summit', *NPR.org*, 23 Sept. Available at: www.npr.org/2019/09/23/763452863/transcript-greta-thunbergs-speech-at-the-u-n-climate-action-summit

Trentmann, F. (2017) *Empire of things: How we became a world of consumers, from the fifteenth century to the twenty-first*. London: Penguin Books.

Turow, J. and McAllister, M.P. (2009) 'General introduction: Thinking critically about advertising and consumer culture', in Turow, J. and McAllister, M.P. (eds.) *The advertising and consumer culture reader*. New York: Routledge, pp. 1–8.

United Nations (2015) *Paris Agreement*. United Nations Framework Convention on Climate Change (UNFCCC). Available at: https://unfccc.int/sites/default/files/english_paris_agreement.pdf

Waterson, J. (2020) 'Guardian to ban advertising from fossil fuel firms', *The Guardian*, 29 Jan. Available at: www.theguardian.com/media/2020/jan/29/guardian-to-ban-advertising-from-fossil-fuel-firms-climate-crisis

Wernick, A. (1991) *Promotional culture: Advertising, ideology, and symbolic expression*. London: SAGE.

West, E. (2022) *Buy now: How Amazon branded convenience and normalized monopoly*. Cambridge, MA: MIT Press.

Index

Printed in the United States
by Baker & Taylor Publisher Services